SCIENCE AND GENESIS

WHAT SCIENCE CAN TELL US ABOUT THE FIRST BOOK IN THE BIBLE

NEVILLE R PIPER

Ark House Press
PO Box 1722, Port Orchard, WA 98366 USA
PO Box 1321, Mona Vale NSW 1660 Australia
PO Box 318 334, West Harbour, Auckland 0661 New Zealand
arkhousepress.com

© Neville R Piper 2020

Unless otherwise stated, all Scriptures are taken from The Holy Bible, English Standard Version© (ESV©) Copyright © 2001 by Crossway, a publishing ministry of Good News Publishers. All rights reserved.

Cataloguing in Publication Data:
Title: Science and Genesis
ISBN: 978-0-6450375-4-8 (pbk)
Subjects: Creation; Christianity;
Other Authors/Contributors: Piper, Neville

Design by initiateagency.com

For Reid:
When you are searching for answers, first be sure you are asking the right questions.

TABLE OF CONTENTS

INTRODUCTION vii

1. THE RISE OF MODERN SCIENCE AND THE SCIENTIFIC METHOD 1
2. IN THE BEGINNING–THE SCIENTIFIC ACCOUNT 8
3. THE EXISTENCE AND NATURE OF GOD 24
4. THE ORIGINS OF THE BOOK OF GENESIS 28
5. IN THE BEGINNING–THE GENESIS ACCOUNTS 36
6. THE ORIGIN OF LIFE ON EARTH 49
7. THE ASCENDANCY OF MAN 62
8. FIRST CONTACT 72
9. CHOICE AND FREE WILL 78
10. ADAM TO NOAH 91
11. SCIENCE AND THE FLOOD 98
12. BIBLICAL SIGNIFICANCE OF THE FLOOD 107

13.	PREDESTINATION AND FREE WILL	114
14.	THE RISE OF CIVILISATIONS	129
15.	ABRAHAM-THE FRIEND OF GOD	136
16.	ISAAC-CHILD OF THE PROMISE	162
17.	HOW JACOB BECAME ISRAEL	168
18.	JOSEPH-FROM SLAVE TO SAVIOUR	179
	CONCLUSION	188

INTRODUCTION

This is not a science book. Nor is it a book about science. It is a book about *Genesis*, the first book in the Jewish *Torah* and the Christian *Holy Bible*, looking at it in the light of our modern scientific understanding of this world and the universe.

By and large, it does not try to justify or explain the wealth of scientific knowledge in the fields of cosmology, astronomy, physics, biology, archaeology and other associated fields of science that are dealt with in these pages. For deeper understanding of these scientific discoveries and theories, you need to go to other sources. My sole concern in these pages is to look at the impact of this scientific knowledge on our understanding of *Genesis*.

From the age of ten, I had no other ambition than to be a scientist, specifically in the field of chemistry. In my final year of schooling I attended an evangelistic meeting, believed the Gospel message and committed myself to Christ. So, as a young Christian believer entering university, I was studying the Bible and science in parallel.

There was no doubting the reality of my continuing experience of God, but I was also being taught that the Bible was the inspired *Word of God* revealed to humankind, written by men inspired by the Holy Spirit of God and protected from error. I soon found myself being confronted with accepted scientific facts and theories that called into question many things written in *Genesis*. As a Christian, was I obliged to take literally all that I read in *Genesis*? If so, how much reliance could I place on the Bible as a whole since every apparent error in Genesis implied that any apparent errors elsewhere in the Bible could throw into doubt the truth and relevance of the whole Bible? On the other hand, if it is not to be taken literally in every respect, where do we draw the line between literalism and symbolism in understanding *Genesis*? I believe that science can contribute much in this area.

In the past century, scientific research has vastly expanded our knowledge in every branch of science and in our understanding of ancient languages and texts. This body of knowledge continues to grow at a rate so fast than it is impossible for any one person to maintain a comprehensive understanding of the whole. I don't pretend to be able to do so. What I have placed here is my own best understanding of relevant current scientific and linguistic knowledge in the hope that the reader will be helped in his or her own mind to come to a confident and sincere resolution of their own beliefs.

1

THE RISE OF MODERN SCIENCE AND THE SCIENTIFIC METHOD

The Scientific Revolution

If we are going to consider the impact of modern science on our understanding of the book of *Genesis*, we first need to understand what modern science is, what it does and how it does it.

For the peoples of the ancient Greco-Roman world, the Earth itself and everything in it and on it were thought to be made up of mixtures of four basic elements—earth, water, fire and air. Beyond the Earth itself, the universe was interpreted as series of concentric crystal spheres revolving around the Earth, one supporting the moon, one supporting the sun, others each supporting one of the known planets and the furthest one supporting the stars. One of the glaring mathematical problems with this model was that the planets, on occasion, would appear to travel backward across the sky over several nights of observation before resuming their normal course. Astronomers called this *retrograde motion*. To account for it, the Greco-Roman astronomer, mathematician and astrologer Claudius Ptolemy of Alexandria in the 2nd Century of the Christian era introduced the concept of 'epicycles' (circles within circles) within some of the planetary spheres. These epicycles allowed apparent retrograde motions of those planets without upsetting the overall operation of the spheres themselves. By and large, belief in the 'Ptolemaic' model of the universe was unchallenged until the middle of the 16th Century.

The emergence of the Renaissance period in the 14th Century saw a breaking away from an almost mindless acceptance of old ideas and customs of the medieval period and led, amongst other things, to an increased reliance on observation and inductive reasoning that led inexorably to what we know as 'science' in all of its areas of operation.

Modern science is widely regarded as beginning with the publication by Nicolaus Copernicus of *De revolutionibus orbium coelestium* (*On the Revolutions of the Heavenly Spheres*) in 1543. In this work, Copernicus compared actual movements of these heavenly bodies relative to the Earth with the expected movements based on the mathematics of spherical orbits centred on the Earth, demonstrating that the long-held belief in an Earth-centric universe could not be true. Although his alternative model of the universe in which Earth, the other planets and the stars all revolved around the Sun was incorrect, it was the first step in unlocking the secrets of the universe. Its true importance lay in its being the first publication in which actual observations of astronomical phenomenon were compared with mathematically derived results from an accepted model and where the model was found wanting.[1]

The next century and a half following on Copernicus' work saw an explosion of activity in physics, mathematics, astronomy, chemistry and

---

[1] Interestingly, Aristarchus of Samos, around 300 BC, had suggested that the universe, including the Earth, circled the Sun. He could offer no evidence for this and the idea was generally disregarded.

biology that became known as the Scientific Revolution. It should be noted that many of the greatest scientific discoveries of that time were made by men of deep faith.

Johannes Kepler (1571–1630 and a contemporary of Galileo) believed that God had created the world according to an intelligible plan that is accessible through reason. In a letter to the Bavarian Chancellor, Hewart von Hohenberg, in April 1599 he wrote:

> *'Those laws [of nature] are within the grasp of the human mind. God wanted us to recognise them by creating us after his own image so that we could share in his own thoughts.'*

Elsewhere he wrote:

> *'I was merely thinking God's thoughts after him.'*

Based on his observations, Kepler derived his *Laws of Planetary Motion* in which the orbits of planets around the Sun are elliptical rather than circular. This causes the planets' speeds to vary, which in turn explained the apparent retrograde movements of planets without recourse to the introduction of 'epicycles'.

Scientist, mathematician and theologian Isaac Newton (1642–1727) derived his three *Laws of Motion* and his *Law of Universal Gravitation*. Using Galileo's experimental data and Kepler's theoretical work on planetary motion, he was able to show that these laws extended beyond the Earth and were instrumental in determining the motions of planets and moons.

Robert Boyle (1627–1691) is widely regarded as the first modern chemist. While mostly remembered for *Boyle's Law* relating to the behaviour of gases under pressure, he passionately rejected the idea that all matter was made up of the basic elements of earth, fire, water and air. In his first book, *The Sceptical Chymist*, published in 1661, he argued for a very early version of atoms, molecules and chemical reactions. He was a devout Anglican who decided against taking Holy Orders because he felt that his theological treatises would have wider impact and acceptance if they were seen to be coming from a layman.

Be they atheist, agnostic or believer, the men of the Scientific Revolution shared a common belief that all matter and physical phenomena are subject to laws that operate uniformly throughout the universe, that these laws are unchangeable and can be expressed mathematically.

Modern scientists frequently use of the word 'Law' in describing and explaining natural phenomena. The reader needs to understand that 'natural' laws are entirely different from 'legal' laws. In legal matters, a 'law' is framed and imposed by a recognised authority and specifies behaviours or obligations that are to be observed by the population to which it applies. Included in that law, specific punishments or sanctions are prescribed for those who fail to obey. In short, a legal law is formulated to bring about a 'desirable' state of affairs in human society.

On the other hand, 'natural' laws set out to describe things as they are. Their purpose is to explain in mathematical terms the rules that determine the 'what?', 'why?' and 'how?' of observed phenomena. They are discovered by men, not made by them.

But how can we be sure that what someone proposes as a 'natural' law is true, and not simply some idea of how the person thinks things should be? The answer to this is that, over time, scientists have developed a rigorous validation process referred to as the *Scientific Method*.

The Scientific Method

The following is a brief general summary of the Scientific Method. It needs to be understood that the variety of scientific disciplines and extreme difficulty of measurements on scales as diverse as astronomical and sub-atomic environments require variations in approach. Nevertheless, it should be clearly understood that the overall integrity of the methodology remains valid.

Observation and Hypothesis

The Scientific Method generally starts with an observation of some natural phenomenon—things always fall 'down', never 'up'; some things float in water while others sink—prompting the observer to ask 'Why is this so?' Most people might answer 'Because God wills it' or 'Because that's the way it is' and never go beyond that thought. Men of science look for the underlying rules that govern a phenomenon. This generally begins with the formulation of a hypothesis (at this stage probably something of an educated guess). A hypothesis tries to identify a quantitative 'rule' that explains how a specific factor could have brought about or influenced the observed behaviour. Where multiple possible contributing factors are identified, the influence of each factor needs to be quantified as a separate 'rule'. Only after the individual contribution of each factor has been determined can it be possible to combine them into a more complex 'rule'.

To be useful, a hypothesis must be able to predict results that should be found when tested in accordance with the 'rule'.

Experimentation and Conclusion

Once a hypothesis is proposed, the next step involves designing experiments in which it will be possible to test actual results against those predicted. It is important that planned experiments identify and document the conditions under which each experiment is to be conducted. A critical component of the Scientific Method is that it should be repeatable by others and that they should be able to get the same results.

If the results of an experiment are not those that were predicted, the hypothesis is wrong either in its entirety or in part or other unidentified factors are affecting the results. The hypothesis must then be rejected or modified. If, time after time, actual results agree with those predicted, the hypothesis may be accepted as a 'theory'.

It is important to understand that a theory is not a statement of proven fact; rather, it is a statement of possible fact. One of the great strengths of the Scientific Method is the obligation on scientists to review, challenge and try to disprove the theory under the widest possible set of circumstances or to find an alternative mechanism that better agrees with the experimental results. Only when all attempts to disprove the theory fail can the theory come to be regarded as a scientific 'law' which accurately explains the phenomenon.

In point of fact, no scientific 'law' can really be regarded as absolutely true because there may be some circumstances yet undiscovered under which the 'law' might fail or need to be modified to take account of hitherto unknown factors. Nevertheless, laws such as those of gravitation and aerodynamics have been found to be sufficiently reliable as to allow us build and successfully fly heavier-than-air aircraft around the world, to land men on the moon and return them to earth and to send spacecraft with scientific equipment to visit every planet in our solar system.

Truths do not contradict. If two 'truths' appear to be in conflict, one or both must be wrong or other unidentified factors must be in play. In such cases, both theories must be revaluated and the apparent differences resolved. In the following chapters, we need to be mindful of the differences between hypothesis, theory and law, between what is and what appears to be, between statement and interpretation.

Our understanding of the Earth and the universe, the development and application of our various technologies and our advances in medical science and chemistry have been built on a brick-by-brick, line-by-line application of the Scientific Method in investigating the world around us. We don't yet know all of the answers (and maybe not even all of the questions), but our extensive knowledge, if not complete, should be sufficient to look at and interpret the validity, meaning and purpose of the *Genesis* stories.

2

IN THE BEGINNING – THE SCIENTIFIC ACCOUNT

Discovering the Universe

That the whole universe had a beginning is something on which the vast majority of modern scientists and the book of *Genesis* agree. But things were not always thus.

Initially, modern scientific theory assumed that the universe is in a 'steady state', always existing but constantly evolving as new stars formed, grew old and died. It was not until the first half of the 20th Century that the true nature of the universe began to be understood.

As late as the early 1920s, astronomers believed that the universe consisted only of the Milky Way galaxy of which our solar system is a part. Stellar objects observed through telescopes which appeared to be 'cloudy' were assumed to be clouds of interstellar 'dust' within the galaxy. Astronomer Edwin Hubble made detailed observations of a number of these 'clouds' during 1922-1923 and was able to demonstrate that they were actually other entire galaxies beyond the confines of our own. The construction of larger and more accurate telescopes, including the Hubble Space Telescope, have now established that there may be around 100 billion galaxies in the known universe, each possibly consisting of hundreds of billions of stars. The Milky Way galaxy is estimated to consist of more than 200 billion stars. (In scientific terms, a billion is one thousand millions; a trillion is a thousand billions)

Developed during 1907-1915 and published in its final form in 1916, Einstein's General Theory of Relativity set out to explain the laws of gravity in relation to other forces of nature in the operation of the universe. Unfortunately, his original equation implied that the universe might be expanding. Influenced by the common belief that the universe existed in a steady state, he was forced to introduce a *Cosmological Constant* into his equation to maintain the status quo. He did this somewhat unwillingly because he believed that its introduction spoiled the beauty of his equation.

A key conclusion of Einstein's theory was that time and space are inextricably linked in a single matrix—one cannot exist without the other. This implies that time *as we know it* cannot exist independently of this universe. This will be significant in later chapters of this book.

Georges Lemaître was an astronomer, physicist and Catholic priest who is credited as the first to have theorized, in 1925, that the universe is actually expanding. His theory was observationally confirmed by Edwin Hubble in 1929. Hubble's work resulted in the discovery of what is now known as *Hubble's Law*, which states that the further away a galaxy is, the faster it is travelling away from the Earth.

The concept of an expanding universe was very controversial, with many scientists engaged in acrimonious debate. Fred (later Sir Fred) Hoyle of Cambridge University was a fierce supporter of the *Steady State Universe Theory*. In a radio broadcast in 1950, he coined the phrase 'Big Bang' as a derogatory reference to the *Expanding Universe Theory*. This term was rapidly adopted (without the derogatory overtones). As evidence accumulated in favour of the 'Big Bang', Einstein was at last able to remove the hated

Cosmological Constant and restore his equation of General Relativity to its original simple form.

Georges Lemaître is also credited with proposing, in 1931, that the observable universe began with an explosion of a single particle. In 1933, at the California Institute of Technology, Lemaître lectured on his theory before some of the world's greatest scientists of the time. At the end of the lecture, Albert Einstein stood up and declared: *'This is the most beautiful and satisfactory explanation of creation to which I ever listened.'*

In 1948, Ukrainian-American theoretical physicist and cosmologist George Gamow made an estimate of the strength of residual cosmic microwave background radiation (the afterglow of the Big Bang) which would have cooled over billions of years, filling the universe with a radiation about five Fahrenheit degrees above absolute zero. This prediction was substantiated in 1965 by Arno Penzias and Robert Wilson who, while engaged in a totally different area of research, accidentally discovered a cosmic microwave background radiation having a temperature of 2.7 degrees Kelvin (The Kelvin scale measures the temperature in Centigrade degrees above absolute zero which is -274°C. Thus 0°C equals 274°K). This discovery is considered a substantial confirmation of the Big Bang theory.

With an acceptance that the universe had a 'beginning', scientists soon began to calculate the age of the universe. Initial calculations ranged as widely as ten billion to twenty five billion years. Current calculations based on a wide range of models all agree that the age is very close to 13.8 billion years.

The Nature of Matter

To understand the story of creation as modern science interprets it, we need to look more widely at scientific discoveries that have been made in recent years in many fields.

My own field of science was chemistry (in particular, radiochemistry). Regretfully, I must restrict my comments here to only those aspects of the history of chemistry that are relevant to the story of creation.

English scientist John Dalton (1766-1844) proposed the modern theory of atoms. The main points of his atomic theory were:

- Elements are made of extremely small particles called atoms;
- Atoms of a given element are identical in size, mass and other properties;
- Atoms of different elements differ in size, mass and other properties;
- Atoms cannot be subdivided, created or destroyed.

William Prout (1785-1850) first proposed ordering all the elements by their atomic weight as atoms of all elements identified to that time appeared to have a weight that was an exact multiple of the atomic weight of hydrogen. Russian chemist Dmitri Mendeleev (1834-1907) developed what became the basis for the modern periodic table of elements in the 1860s. In his table, Mendeleev correctly placed all 60 elements known at the time within the table and predicted the existence of seven new elements, all of which were subsequently discovered.

It was not until around the beginning of the 20th Century that the indestructibility and non-transmutability of atoms was called into question. In 1897, J.J. Thomson discovered the electron and Henri Becquerel and Pierre and Marie Curie investigate the phenomenon of radioactivity. New Zealand-born scientist Ernest Rutherford discovered the internal structure of the atom, the existence of the proton, classified the different types of radioactivity and carried out the first transmutation of an element to another by bombarding nitrogen atoms with alpha particles. His model of the structure of an atom was one that consisted of a tiny but massive positively-charged nucleus around which negatively-charged electrons orbited, much as the planets orbit around the Sun. To account for the fact that many elements seemed to have a mass greater than could be accounted for by protons alone, in 1920 he suggested that the nucleus of an atom could consist of protons together with other neutrally-charged particles of around the same mass as a proton. He gave this theoretical particle the name *neutron*. The neutron was actually detected and identified in 1932 by James Chadwick. As components of the atom's nucleus, protons and neutrons are referred to as nucleons.

Nuclear Physics and Quantum Mechanics

As scientists delved deeper into the structure of atomic nuclei, two things quickly became apparent:

- The proton and neutron were themselves made up of more basic particles which were given the name of 'quarks' (a name coined by Murray Gell-Mann, one of the two physicists who, in 1964, independently proposed their existence). Subsequently, it was established that quarks themselves were made up of even more basic particles. Today, science recognises a 'Standard Model' that contains seventeen *elementary particles* that make up all the matter in the universe (although a search is being undertaken for additional particles that may help to explain the existence of dark matter—discussed at the end of this chapter);
- The physical *laws* that seem to apply at the astronomical scale or even at our human scale don't apply in the sub-atomic world. Within this micro world, other mechanisms determine the behaviour of matter and energy and are the realm of *Quantum Mechanics*. It is not necessary for our purposes to consider this area very deeply except as noted in the following sections.

An unexpected conclusion to this work was the realisation that these elementary particles were also capable of producing antimatter—positively-charged electrons (positrons), a negatively-charged equivalent to the proton and even anti-neutrons. These have been created under extreme experimental conditions, but cannot exist for long in our universe, since they react with the positive-matter equivalents resulting in the total annihilation of both with the release of a large amount of energy.

The Fundamental Forces of Nature

Prior to the 20th Century, only two 'fundamental' forces of nature had been recognised:

- The gravitational force. This is the attractive force that seems to rule the universe from the human to the inter-galactic levels; and
- The electromagnetic force. This force acts between the electrically charged atomic nuclei and electrons of the atom and explains all forms of chemical phenomena, including the production of energy for the use of our bodies from the food we eat. The electromagnetic force acting in a medium also produces electromagnetic 'waves'. The frequency of oscillation of these waves varies according to the manner in which they are produced and the electromagnetic spectrum of these waves extends from radio waves, through microwaves, infra-red, visible light, ultraviolet and x-rays to gamma rays.

Studies of atomic structure led to the recognition of two new forces:

- The strong nuclear force, which binds quarks to form nucleons, and binds nucleons to form nuclei; and

- The weak nuclear force, which binds to all known particles in the Standard Model, and causes (or leads to) certain forms of radioactive decay.

Mass-Energy Equivalence

Einstein's Special Theory of Relativity, published in 1905 and separate from his General Theory, introduced to the world the concept that matter and energy could be converted from one to the other. The equation $E = mc^2$, in which energy and mass are related by the speed of light multiplied by itself, quantified that relationship and suggested that an incredible amount of energy lay within a very small amount of matter.

As a practical example of this, when the atomic bomb *Fat Boy* was dropped on Nagasaki in 1945, it contained 6.15 kilograms of plutonium. The nuclear blast that destroyed the city came from the total destruction of less than a gram of that material. If a kilogram of antimatter was allowed to interact with an equivalent amount of matter, the energy released by their mutual destruction would be about the same as the total energy needs of the United States of America for one day.

The significance of this mass-energy equivalence is enormous. I have already mentioned that we believe there are at least 100 billion galaxies in the universe and that they probably average over 100 billion stars in each. Even taking into account that hydrogen makes up about three quarters of the material in the universe and that helium makes up almost all of the rest, it is hard to conceive that the total mass of the universe could have been concentrated into a ball smaller than an atom. However, if all of that mass was converted into energy, it becomes quite conceivable that that energy could have been concentrated into such a pinpoint providing that the temperature was high enough.

Creation of Galaxies and Stars

Immediately after the Big Bang, the expanding universe cooled and energy was transformed into elementary particles and these in turn were converted into hydrogen and helium atoms. (The story is a bit more complicated than that, as we will see when we look at the scientific creation story below.) Individual local concentrations of these elements were subject to gravity, drawing them closer together to form galactic clouds and, within these clouds, accretions of hydrogen and helium formed massive 'proto stars'. As these proto stars collapsed under their own weight (drawing yet more hydrogen and helium into themselves), eventually the density and temperature of the mass became great enough for nuclear fusion to occur spontaneously, turning the object into a full-blown star. In nuclear fusion, two hydrogen nuclei can fuse to become a helium nucleus and, to a smaller extent, hydrogen and helium nuclei can fuse to form nuclei of heavier elements such as carbon. A star will continue to 'burn' as long as there is enough hydrogen in its core to fuel the fusion reaction.

But early theoretical physical models of the Big Bang predicted that matter (and energy) distribution should have been entirely uniform, preventing local concentrations of matter from occurring. Under these conditions, galaxies and stars could never form. Instead, we know that the distribution of matter is 'lumpy' (or more accurately, somewhat like soapy water within foam where air within the foam represents 'empty' space). So how did the universe become so non-uniform?

In 1979, American cosmologist Alan Guth hypothesized that this *anisotropy* (non-uniformity) of the universe originated early in the first second of creation, when a period of 'inflation' saw the universe undergo a short period of very rapid increase in its rate of expansion (with accompanying minute effects on its uniformity). It is not appropriate here to explain the cause of that expansion; suffice to say that within a billionth of a trillionth of a trillionth (10^{-33}) of a second, the universe expanded by at least 100 trillion trillion (10^{26}) times—from a trillionth of the size of a proton (10^{-25} centimeters) to at least 10 meters across. Inflation ceased as the expanding universe cooled and energy began to 'condense' into matter and expansion of the universe dropped to a more 'leisurely' pace.

While the Inflation Theory successfully accounted for most of the earlier problems with the Big Bang Theory, the experimental evidence to support it was sadly deficient. However, in 1992, American astrophysicist and cosmologist George Fitzgerald Smoot III and his team, used data from the Cosmic Background Explorer (COBE) satellite to identify and map the tiny variations in cosmic microwave background ('wrinkles in time') that confirmed the Big Bang theory (and Inflation). For this he was awarded the Nobel Prize in Physics in 2006. According to the Nobel Prize committee, 'the COBE project can also be regarded as the starting point for cosmology as a precision science.' Stephen Hawking's own assessment of the discovery was that it was 'the scientific discovery of the century, if not of all time'.

Creation of the Elements

Back in 1948, in the same research paper in which he predicted the existence of a cosmic ray background left over from the Big Bang (see the first part of this chapter), George Gamow and his student Ralph Alpher established that the present levels of hydrogen and helium in the universe (which between them make up over 99% of all matter) could be largely explained by reactions that occurred during the Big Bang. This lent further theoretical support to the Big Bang theory. Unfortunately, the elements heavier than helium could not be accounted for in any form of the Big Bang.

It was left to Fred Hoyle, that fierce opponent of the Big Bang theory, to explain how the heavier elements were created. In 1946, he proposed that these elements were produced in stars by *nucleosynthesis*. This process required the immense temperatures and pressures within stars to fuse hydrogen and helium atoms (*nuclear fusion*) to form heavier elements such as carbon. When a star 'dies', the debris, including the heavier elements that have been formed is then available for incorporation into new stars. These heavier elements can be fused into even heavier elements. Hoyle also theorized that other rarer elements could be explained by supernovas, the giant explosions which occasionally occur throughout the universe, whose greater temperatures and pressures would be required to create such elements.

The Story of Creation as Understood by Scientists

Based on the discoveries outlined above, we are now in the position to describe the creation of the universe based on current scientific understanding. The following description is largely based on George Smoot's account in his wonderful book *Wrinkles in Time*[2] and on Charles Lineweaver's article[3] in the first edition of *Newton*, a 'graphic science' magazine published by Australian Geographic Pty Ltd.

In the beginning, space and time were rolled into one as space-time 'foam'. Somewhere around a ten-millionth of a trillionth of a trillionth of a trillionth (10^{-43}) or as late as a millionth of a trillionth of a trillionth of a trillionth (10^{-42}) of a second after the Big Bang, space and time began to

---

[2] Smoot, George and KeayDavidson. *Wrinkles in Time: The Imprint of Creation*. Abacus, 1995.
[3] Lineweaver, Charles. "In the Beginning ... The Origin of the Universe. Part One: Life of the Universe", *Newton*. Sept-Oct, 2000, pp 36-43.

separate to become different coordinates. At that time, the entire universe was about a trillionth of the size of a proton with a temperature of about a hundred million trillion trillion (10^{32}) degrees Centigrade. The three atomic forces of nature—electromagnetism, the strong and weak nuclear forces—were fused as one.

As space expanded, the energy released increased the rate of expansion, introducing the period of inflation so that, by a ten-billionth of a trillionth of a trillionth (10^{-34}) of a second after the Big Bang, the universe had expanded a million trillion trillion (10^{30}) times and its temperature had fallen to below a thousand trillion trillion (10^{27}) degrees. It is important to understand that the universe was not *expanding into* space—it was *creating* its own space. By this time, too, the strong nuclear force had separated and energy was being transformed into elementary particles.

Between a billionth of a trillionth of a trillionth (10^{-33}) and a millionth (10^{-6}) of a second after the Big Bang, a 'soup' of matter and antimatter pairs of particles of all kinds formed from the elementary particles. The weak nuclear force (the only force that can differentiate between matter and antimatter) came into play and may have played a part in ensuring that there was a microscopic difference between the amounts of matter and antimatter being formed. During this time there was a mass mutual annihilation of matter and antimatter, leaving only a small residue of matter that now constitutes the total matter of the universe. At this point, the density of the universe was still greater than the density of an atomic nucleus and was effectively a 'soup' of quarks and electrons so dense that it could be thought of as gigantic nucleus that is the entire universe.

By around a millionth of a second after the Big Bang, with the temperature cooled to a few trillion degrees, the density of the universe dropped to less than the density of an atomic nucleus and the cooling and rarefaction allowed quarks to bind together three at a time to form protons and neutrons.

Between one second and three minutes after the Big Bang the universe had cooled to below ten billion degrees, allowing atomic nuclei to form. When a proton and a neutron collide, they form the nucleus of a deuterium atom (also called 'heavy hydrogen'); when two deuterium nuclei collide, they form the nucleus of a helium atom. At this stage, the universe was still a 'soup' of matter made up of hydrogen and helium nuclei and electrons, dense, opaque and extremely hot.

The universe continued to expand and cool until, about 300,000 years after the Big Bang, it had cooled to 3000°C. At this temperature, electrons and nuclei could begin to combine to form hydrogen and helium atoms. For the first time, the universe began to transform from an opaque 'soup' to become transparent.

It probably took at least one billion years for the first gravitational accretions of hydrogen to become dense enough to form galaxies and then stars. Over billions of years, generations of stars have come and gone, building the multiplicity of elements that have gone to make up our world and the worlds that orbit countless stars in our Milky Way galaxy and in the other hundred billion galaxies. Our own solar system is relatively young. Our own Earth probably was born as part of the creation of our own solar system. Scientists have determined the age of Earth as about four and a half billion years.

Things We Still Don't Know About the Creation Process

The picture of creation that science has painted is both graphic and detailed. However, it has not answered all of our questions and has generated some new ones. In the following, I try to list some of these questions with my understanding of the answers—as far as they are known.

What happened in the first hundred-millionth of a trillionth of a trillionth of a trillionth of a second after the Big Bang?

This question could alternatively be expressed as 'Did this unbelievably minute interval actually exist?' The mathematics that defines the timing of the whole creation process is quite beyond my own understanding, so I have to depend on the theoretical physicists in their calculations and conclusions. They hypothesise that at this stage in the creation process time and space did not exist as separate entities; instead they were part of a space-time 'foam'. In fact, since time and space did not actually exist as such prior to that point in the process, the time interval is essentially theoretical and allows for the appearance of the 'foam' and for it to start 'unpacking'.

What caused the Big Bang?

This is a far more important question but one for which science cannot provide a satisfactory answer.

One theory is that 'outside' of this universe is a 'state' of immense energy at an incredibly high temperature and that in some way a part of this was 'pinched off' to form our universe. This is a highly unsatisfactory explanation because firstly it fails to explain how such a high energy-high temperature could exist or be maintained and contained. It also requires some 'mechanism' by which part of this state could be 'pinched off'.

A second theory suggests that the universe 'tunnelled' into existence from nothing. This is a more acceptable theory for many scientists because it does not require a pre-existing entity and 'tunnelling' is an accepted process within the field of quantum mechanics. At this stage it still needs to identify 'what' it has tunnelled through and how 'nothing' could result in the huge energy and temperature with which it appeared.

Regardless of the theory involved, the greatest problem that scientists face is that we are trapped within our space-time universe and can have no concept of what lies 'outside' it. As much as we might theorize, we currently have no way (or perhaps even prospect) of being able to verify any extra-universal theory experimentally.

A similar situation applies when we ask whether there could be other universes outside our own. The fact that our universe exists seems to imply that other universes could (and maybe should) exist. This is something we almost assuredly will never know and will certainly never concern us. Even if they existed, we do not know what form they might take. The manner in which expansion and cooling within our universe led to the formation of elementary particles and subsequently to the formation of galaxies and stars (and us!), does not mean that every (or even any) other universe similarly formed would develop in the same way.

Dark Matter and Dark Energy

It may appear from what I have written that scientists are confident about the completeness and accuracy of the creation process as described above. This is not strictly true. There are still some significant phenomena that remain unexplained.

In 1975, American astronomer Vera Ruben showed that stars on the fringes of galaxies moved as fast as stars elsewhere in the galaxy but, instead of spinning off and away from the galaxy, they remained in their orbit. This implied that some unrecognised extra mass in the galaxy was at play, holding the galaxy together. This unidentified mass was termed 'Dark Matter'. Calculations suggest that Dark Matter outweighs the amount of known matter in the universe by about ninety times.

Towards the end of the 20th Century, scientists believed that the universe would continue to expand at slower and slower speeds until either it became a static, cold, dead place or that it would stop expanding and, under the influence of gravity, eventually collapse back into itself (the 'Big Crunch'). American astrophysicist Saul Perlmutter studied a class of supernovae referred to as 'standard candles' (because they all emit the same intensity of light) to determine their distances from Earth by measuring their relative brightness. In 1998, as a result of these observations, he showed that in fact the rate of expansion of the universe was increasing (and was awarded a Nobel Prize for his work).

It appears that rather than the energy density of the universe reducing as the universe expands, the energy density remains constant. This implies that 'new' energy is being created or released. According to the First Law of Thermodynamics, energy cannot be created or lost—it can only be transformed from one type of energy to another. If so, then this energy must have been transformed from some unknown form of energy. Science cannot identify the source or nature of this energy and scientists have given it the name 'Dark Energy'.

I am nowhere near qualified or equipped to be able investigate these matters, but I am strongly inclined to the idea that there may be a fifth natural force yet to be discovered. The weak and strong nuclear forces are very strong forces that operate on the nuclear and molecular scales respectively, but their range is limited to atomic dimensions. Electromagnetism is a very much weaker force but can have an effective range of perhaps tens of metres. Gravity is weaker still, but it effectiveness extends over astronomical distances and seems (at least until now) to rule the whole universe. I can conceive of a new force, weaker to the point of near undetectability, which operates at intergalactic distances where even gravity might meet its match and acts to repel rather than to attract.

As better ways of estimating the amount of Dark Matter were developed, it became clear that the total amount of such matter that could exist was too small to account for the way the universe worked; there was a missing component that amounted to about seventy percent of the total energy-mass requirements. Perhaps it is a coincidence, but that missing seventy percent is remarkably close to the amount of Dark Energy equivalent calculated to exist.

So the race is on to discover the nature of both Dark Matter and Dark Energy. It remains to be seen as to how they will impact our understanding of the universe.

And Now?

We know that we exist. We know that our universe exists. We know that the universe as it is had a beginning. But we don't know how it came into existence. It is as difficult to imagine that it came into existence as the outcome of some statistically improbable event within a chaotic firmament that is beyond our ability to ever comprehend as it is to imagine that there is (or was) some supreme power that brought it about. What creative power lies beyond the physical universe?

Scientists are continually increasing our knowledge of the universe and of our place within it. We find ourselves encased in a universe which, from our internal point of view, is 'expanding', but expanding into what? Does 'expansion' have any meaning outside the reality of this space-time 'bubble'? Science has no way of ever knowing. Any knowledge of what is 'beyond', *if it is at all relevant*, must come from another source.

Before we look at the Biblical account of creation, we should first look at what we can know about a possible *Creator* of the universe.

3

THE EXISTENCE AND NATURE OF GOD

In the New Testament, Paul tells us in his letter to the Romans, chapter 1, verses 18-20, that what can be known about God from the physical universe are two things, and only two things:

- God's immense power and
- God's divine nature (that is, his 'separateness' from the universe as we know it).

These are precisely what scientists would agree can be known about a *Creator* from the universe as we have come to understand it.

One of the most widely accepted principles of science is that of Occam's razor. It is a rule of thumb, not a scientific proof. Occam's razor is attributed to the 14th-Century English logician, theologian and Franciscan friar, Father William of Ockham (d'Occam), although the principle was familiar long before that. A popular misinterpretation of the razor says that 'the simplest explanation is the most likely one'. A more accurate statement of the razor is that 'the most likely explanation is the one that explains the observed facts while making the least number of assumptions'.

As we have seen in the preceding chapter, apart from the thorny question of its origin, science is fairly confident that most questions have been answered and we are well on the way to explaining the universe in solely physical terms. On the basis of Occam's razor then, it seems logical to conclude that there is no need of a God. When the famous French scientist Pierre-Simon Laplace presented a copy of his *Celestial Mechanics* to Napoleon, the emperor asked Laplace why his book on the nature of the universe contained no mention of its Creator. 'Sire', he replied, 'I have no need of that hypothesis.'

Any wonder then that so many are happy to reject the existence of God—in terms of modern scientific knowledge of the universe, except perhaps for its origin, God is unnecessary. However, in itself, this does not prove the non-existence of God and we would do well to look a little deeper.

What kind of a God is God?

If there is (or was) a God who created the universe, what kind of a God could 'he' be?

As we have seen above, all that we can learn about God from the physical universe is his enormous power and his independence (otherness) from the universe he created. To go further, we need to make a series of reasonable assumptions (Remember that Occam's razor only tries to reduce assumptions to the minimum required to provide a complete explanation of the observable facts). These assumptions can then be tested to strengthen or weaken the hypothesis:

1. **Is God an intelligent being or simply an unreasoning force or accidental event?** If an unreasoning force or accidental event arising from some pre-existing firmament, the existence of God is irrelevant to us (except as an explanation of the origin of the uni-

verse). On the other hand, if God is an intelligent being, we have to consider another question:

2. **Is God, as an intelligent being, concerned with and interested in that creation?** If 'God' is not interested in this creation, creating it then moving on to some other project, then God should be as irrelevant to us as we are to God. But if God is interested in this creation, we are faced with another question:

3. **Is God only interested in us as an observer, with no interest in controlling the direction that this creation takes or in becoming involved in this creation?** Once again, if God's interest is pretty much that of someone observing an anthill, then God remains irrelevant to our lives. However, if God's interest extends well beyond this, we have to ask a final question:

4. **Is God concerned with guiding and communicating with intelligent life forms within the universe and to what end?** At this stage, we have largely eliminated any reason we might have for regarding God as irrelevant and we need to be looking for answers to what are those plans and purposes for the universe and everything and everyone within it and what are the implications for us.

It seems very reasonable to me (and to many others) that, if there is a God who is concerned with us and wants to establish some relationship with us, there will be some evidence of this, if not overtly in the way the universe operates then in other observations that must be taken into account when applying Occam's razor. We need to consider what kinds of evidence this might be and whether we can find such evidence.

Research into ancient texts can find accounts of direct communication from a divine figure or non-human delegated messenger ('angel') that originate outside of this world and appearing to mankind only in the four monotheistic religions of Judaism, Christianity, Islam and Zoroastrianism. The first three of these have a common basis in the Old Testament.

Zoroaster, living most likely sometime between 1500 and 1000 BC claimed to have had an experience with a shining *Being* and subsequently received further revelations from similar *Beings* in visions. Such 'revelations' did not extend beyond his lifetime.

In comparison, the Old Testament describes multiple experiences of God by many people including a number of first-hand accounts. So if we are to find any continuing historical account of an external God revealing himself to the peoples of this world, we should reasonably expect that we will find this beginning within the Old Testament and probably within the *Book of Genesis*.

> Readers will note that throughout this book I use 'he' and 'him' when referring to God. In Genesis 1:26-27, we are told that God made humankind in God's own image and made them male and female. Therefore, whatever constituted being made in the image of God applied equally to both male and female. Using 'it' to refer to God can be confusing if we are talking about an intelligent personality. Because the man has traditionally been held as the authority figure from hunter-gatherer times until recently, it has been traditional to refer to God as if he was male and it is that common convention that I adopt throughout this book. This in no way is intended to categorise God as male or to belittle women by inference.
>
> I would also point out that Jesus consistently spoke of God as "my Father", "our Father" and "your Father". I do not question his authority and greater knowledge in this matter.

4

THE ORIGINS OF THE BOOK OF GENESIS

Before we look at what the *Book of Genesis* might have to say, we should first ask where this book came from, how it came to survive down through the ages and how reliable are the copies that are currently available to us. Here science, especially the sciences of archaeology and linguistics have much to teach us in a process commonly referred to as *Biblical Criticism* and which can be further broken down into 'Higher' and 'Lower' Biblical Criticism.

'Higher' Biblical Criticism

Until the advent of the so-called '*Age of Enlightenment*' of the 17th to 19th Centuries, no-one seriously questioned the belief that *Genesis* had been written by Moses during the 40-year wanderings of the Children of Israel in the desert regions of the Middle East. But, as part of a general reassessment of the whole Bible, a number of *Enlightenment* thinkers (starting with Benedict Spinoza (1632-1677) began to question Moses' authorship of the *Pentateuch* (the books of *Genesis, Exodus, Leviticus, Numbers* and *Deuteronomy*) when they noticed that some sections of the ancient Hebrew texts contained language forms that reflected language usages dating from different historical periods. Their conclusion was that *Genesis* had been assembled from a variety of sources that had themselves been written at different and more recent times.

Higher criticism recognises that all ancient stories began as oral creations but, when put into writing at a later date, were likely to reflect the literary forms, colloquialisms and social context of the age in which the writer lived. Over the last one hundred years, archaeologists have uncovered huge amounts of written material in the form of clay tablets, papyrus and parchment document fragments and texts carved into rock and marble, many of which can be confidently dated to historical periods and geographical locations. Other texts for which a historical period and location are uncertain can then have their characteristic use of language forms, colloquialisms and social context matched with similar characteristics of dated texts and can thus be provisionally assigned to the same historical period and location.

The field of Higher Biblical Criticism has been and continues to be hotly contested (after all, academic reputations are generally not to be gained by agreeing with the opinions of your predecessors or rivals). However, one of the longest-held theories in Higher Criticism was mapped out by Julius Wellhausen (1844-1918), who suggested that *Genesis* had been assembled from four separate and overlapping documents referred to as the *Yahwist* or *J* document (currently thought to have been written in the 10th-9th Centuries BC), the *Elohist* or *E* document (currently thought to have been written in the 9th Century BC), the *Priestly* or *P* document (currently thought to have been written in the 6th-5th Century BC) and the *Deuteronomist* or *D* document (currently thought to have been written at some time in the 7th-5th Centuries BC). In one form or another, this theory has persisted as the *20th Century Documentary Theory* until the present day, although arguments still go on as to whether the J document ever existed as a separate document or whether the *D* document is actually a composite of two earlier documents

Dtr1 and *Dtr2*. It is assumed that the final version of the *Pentateuch* was created from these documents by one or more unknown *Redactors* (editors), perhaps as late as the 5th Century BC. The *D* document is thought to have contributed nothing to the final version of *Genesis*.

The greatest limitation of Higher Criticism is that, while it may offer some help in dating these proposed earlier *J*, *E*, *P* and *D* documents that may have been used as the basis for the *Book of Genesis* as we now know it, it tells us nothing about the original source materials used in compiling these earlier documents. Given the degree of overlap of stories included in these earlier documents, it is reasonable to assume that the 'authors' of these documents were themselves working from at least fragmentary copies of even earlier '*Genesis*' documents. There is no doubt that *Genesis* was created from multiple sources (even *Genesis* itself makes this clear), but as to who was/were the original compiler/s and who were responsible for later revisions will probably never be satisfactorily determined. It is certain that some editing (and perhaps even enhancement) of these stories has occurred over the generations that the various documents have existed.

For reasons that I go into later in this book, I choose to accept that the first compilation of *Genesis* (whatever it contained), was assembled in the time of Moses, more likely under his or Joshua's supervision rather than by Moses' own hand, but using documents that Moses may have gathered before leaving Egypt as well as documents that may have been passed down through generations of the Hebrews that came out of Egypt with Moses. However, our concern in this book is not with the history of successive versions of document(s) that became the *Book of Genesis* but with:

- scientifically evaluating the likely historicity of the stories found there,

- seeing whether these provide reasonable evidence of a God who wishes to establish a relationship with humankind; and
- evaluating their value in understanding the nature of God and God's purposes in this world.

Critical to all of this is the question of how much trust we can put in the accuracy and completeness of the *Genesis* texts as we have them today. This is the realm of Lower Biblical Criticism.

'Lower' Biblical Criticism

Lower Biblical Criticism is essentially concerned with determining the accuracy and dependability of the existing Bible texts.

Each time copies are made of ancient texts, the copies are expected to be faithful reproductions of the documents from which they are copied. However, this may not always be the case because:

(1) a copy (version) may include deliberate additions, changes or omissions designed to promote the interests and beliefs of the compiler,
(2) the copyist may make deliberate additions or changes to 'help' the reader's understanding (for instance by updating the name of a location to a more modern name for that location), or
(3) mistakes may be made inadvertently by the copyist and the errors are not detected and corrected. Many examples have been found in recent times of individual copies of handwritten Bible texts where it is clear that the eye of the copyist has inadvertently wandered from a word he was copying to an earlier occurrence of the same word on the original page (so that part of the original text is duplicated) or to a later occurrence (so that part of the original text is omitted from the copy). Other instances have been identified

where the copyist's poor penmanship has resulted in a word being entirely changed.

Whatever the reason, subsequent copies will reflect those changes and the value of the text as a copy of the original document is compromised. Over time, changes can accumulate, even to the extent of altering the original meaning or purpose of the source document.

Trying to establish what parts of a text can reliably regarded as 'original' and the nature and impact of textual changes is an extremely complex task because we do not have access to the original document(s). Instead, we must rely on a series of assumptions and tools to determine the likely reliability of the documents that we do have. I am indebted to the prominent Bible scholar, Fr Mitch Pacwa, for an explanation of these 'tools' (and a masterly summary of current thinking on Higher and Lower Biblical Criticism).[4]

In the case of the Jewish Bible (the Christian *Old Testament*), we do not have the original hand-written source document(s); we are entirely dependent on generations of handwritten copies without any way of knowing how faithful and reliable each might have been. Our best hope is to identify mistakes and changes by careful word-by-word comparison of the copies. Where differences between copies are located, two rules are generally applied:

---

[4] Pacwa, Mitchell. "Critical Methods of Scripture", *Deep in History* Conference, 2010 organised by The Coming Home Network. Available on YouTube.
Fr Mitch Pacwa is a Jesuit priest and an accomplished linguist, fluent in several ancient languages (Latin, Koine Greek, Hebrew, Aramaic, and Ugaritic) as well as the modern languages of English, German, Spanish, Polish, Modern Hebrew, Arabic, French, and Italian.

- The oldest copies are regarded as most likely to be more correct since they are closer to the period in which the originals were written and therefore closer to the thought forms of the original writers; it is also probable that fewer errors would have been accumulated to that date.
- Where many copies are similar in age, the likelihood is that the majority of the copies will have been made correctly and the minority with textual differences can be discounted.

Until the middle of the 20th Century, the earliest Ancient Hebrew version we had of the Jewish Bible was assembled by the Masoretes (a group of Jewish scribes and scholars during the 6th—10th Centuries AD) from a collection of earlier texts. The oldest actual physical copy of this Masoretic text still in existence is the *Leningrad Codex* (book) dating from 1008 AD and constitutes the official text of the Jewish Bible.

Complicating matters is the fact that the original Ancient Hebrew alphabet, like most middle-eastern languages of the time, had no vowels. This meant that, in many cases, a written word could have many possible meanings depending on the way it was pronounced and the correct pronunciation had to be inferred from the context. The Masoretes introduced additional vowel symbols to the Ancient Hebrew text that indicated how a word was to be pronounced, thus clarifying its intended meaning. The determination of meaning was usually based on oral tradition amongst the elders and could sometimes be very subjective; in these cases, majority agreement between the Masoretic scholars determined the matter and may possibly have been wrong.

In November 1946, two Bedouin shepherds discovered scrolls housed in jars in a cave near an area known as Qumran. Over the next three months they explored three additional caves, discovering more scrolls in earthenware jars. They retrieved a handful of scrolls and offered to sell them to local dealers in antiquities. Eventually, the documents came to the attention of Dr. John C. Trever, of the American Schools of Oriental Research. Over the next decade, a total of eleven caves were discovered and excavated. Most of the material found consisted of thousands of fragments of larger manuscripts that had deteriorated, but a small number of well-preserved, almost intact manuscripts also survived—fewer than a dozen among those from the Qumran Caves. Altogether, researchers have assembled a collection of 981 different manuscripts, all of which date from the last three centuries BC and the 1st Century AD.

Amongst these manuscripts were Ancient Hebrew copies of every Old Testament book with the exceptions of *Esther* and *Nehemiah*. These now provided at least partial copies of the Old Testament books that pre-dated the *Masoretic* texts by 900-1,200 years. The interesting thing about these partial copies was that they largely confirmed the accuracy of the *Masoretic* texts (but of course without providing any new light on the accuracy of the vowels introduced by the Masoretes).

The other major tool in assessing the accuracy of the Ancient Hebrew texts is examination of early translations of the texts into other languages. The logic behind this approach says that the translators worked at a time much closer to the time the texts were written and, in order to produce an accurate translation, they needed to be familiar with the Jewish understanding of those texts at that time.

The most significant ancient translation of the Jewish Bible (among a large number of translations of Hebrew Bible books) is that of the *Septuagint*, a translation into Koine Greek of the whole Jewish Bible plus other related documents. So-called because it was reputed to have been begun by seventy Jewish scholars in Alexandra (from the Latin *septuāgintā*, meaning 'seventy' and often abbreviated as the Roman numerals LXX), work began in the middle of the 3rd Century BC and continued into the 1st Century AD. The *Septuagint* was the basis for further translations into Latin (the '*Vulgate*') and the Syriac, Armenian, Coptic, Slavonic and Georgian languages. Comparisons between the Masoretic texts and the various Septuagint and 'downstream' translations have established that the current and Greek versions are overwhelmingly accurate copies of any manuscripts that were in existence in the last three centuries BC. The reverence and care with which the Jewish priestly class handled and copied their scriptures prior to that time give us a high level of confidence that a similar concordance existed between the later copies and those that extended back to at least the time of the Babylonian captivity.

Now with an expectation that the text of *Genesis* as we have it is a reasonably accurate copy of the original compilation of 'beginning' stories, we can start to assess their possible historicity and significance.

5

IN THE BEGINNING-THE GENESIS ACCOUNTS

Types of Creation Myths

Stories of the Creation exist all around the world but, as there was no-one there as an eyewitness, all must be regarded as myths. Creation myths can be classified into five types:

- Creation *ex nihilo* - the idea that God (or gods) created the world out of nothing;
- Creation from *chaos*—the idea that God (or gods) created the world from a pre-existing formless expanse or firmament that contains the material with which the created world will be made;
- 'World Parent'—the idea that the world was created as a result of separation of two tightly bound primeval entities or grew from a dismembered part of a single primeval entity, the world parents or parent respectively;
- 'Emergence' where humanity emerges from another world into this one. Strictly speaking, this type of myth only is concerned with the advent of mankind into a world that already existed; and

- 'Earth-diver' which takes one of two forms - the first where God sends a being (usually an animal) into pre-existing primal waters to find material with which to build habitable land, the second where potential beings are suspended in the primordial realm and which awaken to build lands where they will be able to live.

Only the first two these classifications are scientifically acceptable and, in fact, are compatible with the two main theories of the origins of the universe—the 'tunnelling' from nothing and the 'pinching off' of a part of a pre-existing firmament.

The Genesis Accounts of Creation

The reference to 'accounts' in this chapter's title is not accidental. In *Genesis* we are given two separate and, to some extent, conflicting accounts of creation, the first in Genesis 1:1–2:3 and the second in Genesis 2:4–25. The two accounts differ in literary style, names of God and sequence of events. We will look at each of them in turn.

The 'YHWH' Account (Gen. 2:4-25)

This account is so-called because it refers to God by the Tetragrammaton 'YHWH' used throughout the rest of *Genesis* and elsewhere in the Old Testament. It is pronounced in Hebrew as 'Yehovah' and generally translated into English as 'the Lord'.

This account begins with 'These are the generations of the heavens and the earth when they were created, in the day that the LORD God made the earth and the heavens' but ignores the actual creation of the earth and the heavens, starting with an earth that already exists. The text says that the were no plants because rain had not yet fallen, suggesting that seed could have already been in existence awaiting rain before man was created 'from the dust of the ground'. Verse eight suggests that the Garden of Eden was already established when the man was placed there.

Some commentators suggest that this account pictures animals as being created after man and being presented to the man as 'helpers'. However, the text can also be interpreted as God presenting animals that had previously been created, so that this 'difference' from the 'Elohim' account below is probably not significant. There is no doubt that the appearance of the woman is seen as a separate act of creation by God but, unlike the man's creation and the creation of the animals, this was 'out of' the man, not from the dust of the ground.

There is no attempt in this account to provide detail of any of the creation process other than to portray man (and animals) as creatures formed from the constituents of the earth on which they live and as the direct result of God's creative activity. No timescale or sequence of the creative process is provided—the 'day' in verse 4 is clearly intended to the period of creation, not necessarily a 24-hour day. The primary purpose of this account is to introduce the story of the Hebrew people as receivers of the unfolding revelation of God to mankind.

The 'Elohim' Account (Gen 1:1—2:3)

This account takes its name because, unlike the rest of *Genesis*, God is here referred to only as *Elohim*, the Hebrew word for God. While the 'YHWH' account is an integral part of the *Genesis* story, the 'Elohim' account appears to have come from another source. Apart from the name assigned to God, this account is much more structured in form, with creation neatly organized into 'days'. Each of the days of creation commences with 'And God said' and concludes with the formulaic expression 'And there was …'. By the seventh day, all creation exists in its proper sphere, and God rests.

But if this account was indeed copied from another source, the questions arise as to why and when it was attached.

Some commentators suggest that it could have been added during or immediately following the Babylonian captivity as part of a concentrated effort by priests and scholars to provide a 'complete' and 'authoritative' (in accordance with their own views) set of scriptures to support their teachings and authority. I think this is very unlikely for two reasons:

- Editing at that stage would almost certainly have ensured the use of *YHWH* in reference to God in keeping with the rest of *Genesis* and
- The Ten Commandments handed down at Sinai, described in Exodus (the second book in the *Torah* and *Old Testament*), includes the observance of the Sabbath and is specifically stated to be based on the 'seventh day of creation' as a day rest for God. No other part of *Genesis* alludes to the days of creation.

Back in Chapter 4, I stated my belief that *Genesis* was compiled in the time of Moses. If I am correct, we need to consider just what sources were available as a basis for the book.

The first of these would certainly have been written accounts of the Hebrew forefathers that had been in the possession of and carefully preserved by the Hebrew people. History might not have been your favourite subject at school but most of us have an inherent desire to know where and who we came from. This was particularly important to the Hebrew tribes because of their need to understand their unique place in the world as the *Chosen People of God* (We will speak more of the origins of these stories in Chapter 10). But why preface the '*YHWH*' story with the '*Elohim*' account and where did it come from?

Moses grew up in Pharaoh's court and was trained in all of the knowledge of the Egyptians, and this would have included their religion. Moses clearly knew himself to be a Hebrew and identified with them as his people. At that time he almost certainly accepted their one true god as his also.

When the author was compiling the *Book of Genesis*, it must have been clear that the 'YHWH' account of the story of the Hebrew people leaves untouched the whole issue of the creation of the universe and our world. This is an issue the author may have wrestled with. No doubt Moses himself had full access to the creation myths of both the Sumerian and the Egyptian peoples during his education, but these essentially involved many gods squabbling over an already half-formed world and were completely antagonistic to the Hebrew understanding and experience of one God. It isn't clear where the 'Elohim' account originated; from the language and form it seems unlikely that Moses or any other Hebrew concocted it. Conceivably it could have been given to Moses by revelation from God but there is nothing to suggest this or why he would choose to not to use *YHWH* for the name of God. The most reasonable suggestion is that Moses found this account amongst the document collections of the Egyptian court (or someone in the Babylonian captivity found it in a Babylonian library) and regarded it as most in keeping with the God in whom he believed.

The 'Elohim' account relates God to the whole universe as both its creator and absolute ruler and to mankind as God's ultimate creation. We are left in no doubt as to the mighty power of the God with which we are dealing or the unique nature of humankind who bear the '*image of God*' within ourselves.

The Purpose of the Genesis Creation Accounts

From a scientific point of view, both accounts fall well short of reality as we currently understand it. That is hardly surprising since neither are 'eyewitness' accounts. For that matter, no one was around at the time to confirm our current scientific creation scenario; it is only a theoretical account based on our current knowledge and seems to be supported by our discoveries to date.

If we wish to recognise the Bible as the inspired word of God and accept that the *Genesis* stories are a valid part of that revelation, we need to consider the limitations of the world into which it was delivered. God's earliest interactions with mankind were to people who were sparsely scattered in small groups and who were socially and scientifically primitive. If God had chosen to give mankind a true account of creation at that time, it would have been incomprehensible to them (it's hard enough for scientists today!). We need to be less concerned with the accuracy of the stories and deeply concerned with the principles that God was teaching mankind through their symbolism. In the creation accounts, we are given a picture of a mighty God, creator and ruler of the universe, who raised humankind 'in his image' and has chosen to be known by and to associate with them.

Unpacking the Genesis Accounts

In the 'Elohim' account we can infer two kinds of creation:

'In the beginning God created the heavens and the earth.' This verse very much implies a creation *ex nihilo*, but the following verse states that what was created was essentially formless, empty and dark. From the text it can be argued that this initial creation was a separate event and could have occurred at any time prior to the first 'day' of creation when God began to bring order out of the state of *chaos*.

This fits well with the scientific account in which initially the universe was formless and essentially empty. Light did not appear until somewhere between 200 million and a billion years after the Big Bang as the first stars formed within the immense pressures and temperatures of the hydrogen clouds that were forming the galaxies. However, we should not interpret this as suggesting that the author of the account had some scientific insights into this stage of creation.

The 'days' of creation are no less problematical. The Hebrew word *yowm*, here translated as 'day' has a number of meanings associated with time:

- Daytime (as opposed to night-time)
- A precise period of 24 hours
- A general period of time as in a *working day* or *a day's journey*
- A lifetime (as in '*all the days of Enos were nine hundred and five years: and he died.*')
- A specific time or event (as in *the Day of the Lord*)
- A specific period or age (as '*in those days*')

Some commentators have suggested that one of the broader translations might have been intended, but the text itself is quite clear in indicating a 24 hour period by referring to *the evening and the morning*. The fact that the Sun was not created until the fourth day, making the existence of mornings and evenings meaningless before then, may seem an obvious error to us but misses the main point—the lesson was that the creation of the world was a planned and structured process. The six days of creation lay out a hierarchy of creation:

- Day 1—Creation of light, the prerequisite to all life and separation of light from darkness to form day and night (the first basis of time measurement).
- Day 2—Separation of the waters between those waters beneath and the clouds above and creation of the atmosphere.
- Day 3—Separation of the land from the waters. This also included the creation of plant life, which was probably not understood to

be a life form as such, but part of the sustenance provided by the world to its inhabitants (although it is clear that the peoples of the time that the *Elohim* account was first recorded understood the nature of seed-based propagation).
- Day 4—Creation of the heavenly bodies allowing a second basis for measuring time into seasons and years based the movements of the heavenly bodies—primarily the Sun.

These first four 'days' set the stage on which created life was to dwell, while days 5 and 6 are concerned with the population of that stage.

- Day 5—Population of the seas and skies with life.
- Day 6—Population of the land with living creatures and, as a distinct creation, mankind—male and female—in God's 'own image'.

> *[25] And God made the beasts of the earth according to their kinds and the livestock according to their kinds, and everything that creeps on the ground according to its kind. And God saw that it was good.*

It is interesting that this account differentiates between 'livestock' and 'wild animals', once more indicating its origin come from a time when man had already begun to domesticate some animals.

> *[26] Then God said, "Let us make man in our image, after our likeness. And let them have dominion over the fish of the sea and over the birds of the heavens and over the livestock and over all the earth and over every creeping thing that creeps on the earth."*
> *[27] So God created man in his own image, in the image of God he created him; male and female he created them..*

> *²⁸ And God blessed them. And God said to them, "Be fruitful and multiply and fill the earth and subdue it, and have dominion over the fish of the sea and over the birds of the heavens and over every living thing that moves on the earth."*

The concept in Gen 1:26 of mankind being made 'in the image' of God is dealt with in more detail in the Chapter 7, but the inclusion of both male and female in verse 27 makes it clear that God's image was not to be thought of in purely male terms. Moreover, being made in the image of God was specifically to fit them for the responsibility of 'having dominion' over all living things (verse 28) and (in Gen 2:15) to work and care for the land.

> *²⁹ And God said, "Behold, I have given you every plant yielding seed that is on the face of all the earth, and every tree with seed in its fruit. You shall have them for food.*
> *³⁰ And to every beast of the earth and to every bird of the heavens and to everything that creeps on the earth, everything that has the breath of life, I have given every green plant for food." And it was so.*

These verses suggest that all land-based creatures and man were initially to be vegetarian. Science knows this not to have been true, but it fits with the creation account of the third day, where plant life is regarded as that part of the created land intended to sustain the animal and human life to come.

The 'YHWH' account in Genesis 2 is primarily concerned with mankind and specifically with mankind's place within the purposes of God.

> *⁷ then the LORD God formed the man of dust from the ground and breathed into his nostrils the breath of life, and the man became a living creature.*

The Hebrew phrase *neshamah chay*, translated as 'breath of life', should not be thought of as relating to 'the image of God' of Gen 1. It is used consistently throughout the Old Testament in relation to any human or animal life (anything that draws breath).

> *¹⁸ Then the LORD God said, "It is not good that the man should be alone; I will make him a helper fit for him."*
> *¹⁹ Now out of the ground the LORD God had formed every beast of the field and every bird of the heavens and brought them to the man to see what he would call them. And whatever the man called every living creature, that was its name.*
> *²⁰ The man gave names to all livestock and to the birds of the heavens and to every beast of the field. But for Adam there was not found a helper fit for him.*
> *²¹ So the LORD God caused a deep sleep to fall upon the man, and while he slept took one of his ribs and closed up its place with flesh.*
>
> *²² And the rib that the LORD God had taken from the man he made into a woman and brought her to the man.*
> *²³ Then the man said, "This at last is bone of my bones and flesh of my flesh; she shall be called Woman, because she was taken out of Man."*
> *²⁴ Therefore a man shall leave his father and his mother and hold fast to his wife, and they shall become one flesh.*

Some commentators use these verses to justify women's subservience to man because she came 'out of' man. Even the apostle Paul is guilty of this. However, the primary purpose of the passage is to emphasize that, if God's plan is for mankind to multiply on the earth and to dominate and care for it, a man cannot do it by himself or in companionship only with non-human life forms (or for that matter, with another man). Instead, God's plan is to be fulfilled through stable sexual relationships between males and females of the same genus (literally 'of the one flesh') to create progressive generations.

We will continue our consideration of the *Genesis* story in Chapter 8.

The Purpose and Plan of God

We have made the point in Chapter 3 that there is no point in believing in or seeking to communicate with a Creator God unless it is first clear that God is interested in establishing a relationship with us. Most of my readers will already believe that he has indicated this in many ways and at many times and that the earliest communications are recorded in *Genesis*. It is here that we begin to understand the nature of God and his purposes.

So, before we leave the subject of the Creation, we would do well to think about its purpose (if we agree that there is a purpose).

We have talked about the immensity of the universe in terms of both its age and size. If God's purpose was purely to raise up and communicate with mankind, why go to such an effort? Surely he could have done exactly what Genesis 1 describes and saved a great deal of time and energy.

Of course, we need to understand that God has every right to do what he wants in the way he wants to do it. We are not in a position to judge. The story of Job teaches us that we have neither the knowledge nor the wisdom to challenge God's purposes; we can only submit to his will and trust that he has our best interests in mind.

Having recognised that, we might still consider some factors that may have been significant in the mind of God.

The first of these should surely be to demonstrate the vastness of his power, not just to the most primitive peoples, but also to the scientifically competent peoples to come later.

Secondly, God as he is depicted seems to make it abundantly clear that he wants a relationship that involves a willing participation on our part. This means that we cannot be constrained within a narrow reality where awareness of God's presence dominates. There is no willingness where rejection is not a viable option. We need a universe in which God's deniability is possible.

We live in an imperfect universe and an imperfect world; natural disasters—volcanic eruptions, floods, bushfires, famine and disease kill millions every year. In the face of such challenges, mankind has been forced to use his curiosity and intelligence to understand the world and overcome every difficulty. I believe, as do many, that this was very much part of God's plan to expand our understanding of ourselves and of God himself. It is all part of bringing us to maturity and fitting us to engage in real relationship with him.

And a closing thought:

Despite our growing knowledge of the size of the universe, we still think of ourselves as its centre and the sole object of God's interest and love. Why should we think that we are alone in the universe? What if God is raising up creatures to seek and worship him all over the universe? We will probably never know the answer this side of eternity.

Our knowledge of God is limited to what he has chosen to reveal to us. Our rational response to this revelation should be to seek the personal relationship with him that he offers.

6

THE ORIGIN OF LIFE ON EARTH

The Theory of Evolution

If the expanding universe was the greatest scientific 'discovery' of the 20th Century, then surely the evolution of life from single cell life to the complex world of today's flora and fauna, including mankind, was the greatest scientific 'discovery' of the 19th Century.

In November 1859, eminent scientist Charles Darwin published *On the Origin of Species by Means of Natural Selection*. This book caused a furore around the world because it proposed that all forms of life on earth evolved over geologic time from a single life form.

Darwin was not the first to suggest this. For nearly a century before that time, some scientists (including Darwin's grandfather, Erasmus Darwin) had suggested the possibility of species transmuting into newer species over many thousands or even millions of years. (Science had already begun to realize that the earth was very old—far older than the six thousand years that Bishop Ussher had calculated as the age of the world based on the Biblical account.) The problem that they faced was a lack of a mechanism by which this could be accomplished.

In his book, Darwin introduced the theory that populations evolve over the course of many generations through a process of natural selection. His theory was based on the accepted fact that individuals within a life form can exhibit significant natural differences from other members of its generation. Some of these differences are disadvantageous and the individual will be less successful in the competition for limited resources that it needs to survive or that make it less able to survive in the environment in which it finds itself. That individual will either die or be less successful in passing on its inherited traits. On the other hand, if an individual had traits that gave it an advantage in competing for resources or that enabled it to better survive in a changing environment, that individual had a better chance of survival and of passing on those traits to the next generation. The book presented a mass of well-researched examples across many life forms from many parts of the world that strongly supported the theory. Not only did the theory explain changes within a population over time as their environment changed, it also explained how a life form could branch into different evolutionary directions when different parts of the environment occupied by a life form underwent widely differing changes (as would be the case when some parts of the environment slowly became drier while others became wetter).

Since 1859, advances in geology, biochemistry and palaeontology (in the expansion of the fossil record) have all contributed to further strengthening the theory to the point where, in its modern form, it is wholly or partly accepted by most scientists (Even most scientifically-trained proponents of *Intelligent Design* accept that evolution occurs, if only to a limited extent).

What Should We Make of the Genesis Account of the Creation of Life?

If, as I have suggested earlier, the *Genesis* accounts of the creation of the world and the universe were deliberately attuned to the minds of primitive man, the account of the creation of life was even more so. As the Genesis 2 accounts says:

> [7] *then the LORD God formed the man of dust from the ground and breathed into his nostrils the breath of life, and the man became a living creature.*

As mentioned in Chapter 5, the Hebrew phrase *neshamah chay* (translated as 'breath of life'), is used in relation to any human or animal life (e.g. Gen. 1:30, 6:17 and 7:15). When a human (or animal) is breathing it is clearly alive; when it stops breathing, it is clearly dead.

Thus the concept of God 'breathing' into a man (or animal) is something that early mankind could readily grasp and allowed them to understand that life itself is a gift from God.

What Do We Mean by 'Life'?

Scientists estimate that Earth is currently home to as many as 8.7 million different forms of life, so 'breathing' is no longer adequate as a basis for determining what is 'alive'. Instead, scientists distinguish living organisms from dead organisms and inanimate matter in terms of functions such as *homeostasis, metabolism, growth, reproduction*, and *heredity*.

Homeostasis refers to the ability of the body or a cell to seek and maintain a condition of equilibrium or stability inside itself when dealing with external changes.

Metabolism refers to biochemical processes that occur within a living organism. Metabolism is of two forms: anabolism (the building up of substances) and catabolism (the breaking down of substances). The most easily recognised metabolic process is the breakdown of food and its transformation into energy to maintain life and growth.

Heredity refers to the passing on of genetic factors from one generation to the next.

The Chemistry of Life

In the century and a half since the publication of *The Origin of the Species*, the growth in our understanding of the chemistry of life has revealed it to be so much more complex and fantastical than we ever could have imagined and even now seems to be nothing short of miraculous. The following is only a very simplified description of the biochemical processes involved in the maintenance of life.

For the vast majority of life forms currently in existence on Earth, the biochemistry is based on **deoxyribonucleic acid**, commonly referred to as **DNA**. DNA are complex molecules that contain an organism's genetic code. The code tells the cell what kinds of proteins it needs to make. The code itself is made up of four chemical bases: adenine (A), guanine (G), cytosine (C), and thymine (T). The order in which these bases appear in the DNA molecule and the length of the molecule (and thus the number of bases that make up the code) determines the information available for building and maintaining the organism.

DNA bases pair up with each other, A with T and C with G, to form units called base pairs. Each base is also attached to a sugar molecule and a phosphate molecule. When a base molecule, sugar molecule and a phosphate molecule link together, the combination is called a **nucleotide**. In

DNA, nucleotides join in two long strands that form a spiral called a double helix (something like a helical ladder) where the base pairs form the rungs and the sugar and phosphate molecules form the sidepieces of the ladder.

When cells divide, each new cell needs to have an exact copy of the DNA present in the old cell. To achieve this, an enzyme called *DNA helicase* splits the DNA down the middle by breaking the hydrogen bonds between the base pairs that make up the 'rungs' of the DNA 'ladder', leaving two single strands of genetic material. Then another enzyme, called *DNA polymerase*, builds a new strand onto each of the original strands by matching the bases (A with T and G with C). In this way, each copy of a DNA molecule is made of half of the original molecule and half of new bases and each new molecule is the duplicate of the original.

When DNA is copied, mistakes sometimes occur—these are called **mutations**. There are three main types of mutations:

- *Deletion*, where one or more bases are left out.
- *Substitution*, where a base is substituted for another base in the sequence.
- *Insertion*, where one or more extra bases are inserted into the sequence.

Mutations may be bad or beneficial for the organism or have no significant effect on the cells. 'Bad' mutations are fatal for the organism—the protein made by the new DNA does not work in the way it needs to and the life form dies. On the other hand, a 'good' mutation may produce cells in which the new version of the protein works better, giving the organism an advantage. It is essentially through mutations that evolution moves forward.

Insertion of additional bases within DNA has been particularly important in the development of more complex life forms. The simplest forms of life at present have hundreds of bases within their DNA, whereas human DNA consists of about 3 billion bases. One argument that opponents of evolution raise is that the Second Law of Thermodynamics states that entropy (disorder) of a system always increases and so maintain that complex life forms can only evolve into less complex ones. The fallacy of this argument lies in the fact that base insertions do occur and, if the result is an attribute or attributes that enhance the survivability of the life form, it will succeed in reproducing itself.

The genetic code for an organism does not consist of a single strand of DNA. Rather, there are many separate and quite different sets of strands, each one responsible for a particular function, such as the production of a specific protein. Because the DNA molecules in most organisms are very long and fragile, they are packaged into thread-like structures called ***chromosomes***. Each chromosome is made up of DNA tightly coiled many times around proteins called histones that support its structure and each chromosome has a specific role to play in the development of that organism.

In humans, for instance, each cell normally contains 23 pairs of chromosomes for a total of 46. Twenty-two of these pairs, called *autosomes*, look the same in both males and females. The 23rd pair, the sex chromosomes, differs between males and females. Females have two copies of the X chromosome, while males have one X and one Y chromosome. Occasionally, an individual may be born with three sex chromosomes, two of one type and one of the other. We now recognise these as 'transgender' individuals.

Within chromosomes, sections of DNA that code for a specific function are called ***genes***. Some genes act as instructions to make molecules called proteins. Proteins form structures and also form enzymes. Enzymes do most of the work in cells. Proteins are made out of smaller polypeptides, which in turn are formed from amino acids. To make a protein to do a particular job, the correct amino acids have to be joined up in the correct order.

A gene is the basic unit of heredity. Every person has two copies of each gene, one inherited from each parent. In humans, genes vary in size from a few hundred DNA bases to more than 2 million bases. Humans have about 20,000 genes. More than 99 percent of these genes are the same in all humans; the remainder account for variations between races and differences between individuals. The complete set of DNA belonging to and defining an organism, including all of its genes is referred to as its ***genome***.

While scientists have developed quite a deep understanding of the biochemical processes that sustain life, there is still so much to learn. As our knowledge of the genetic make-up of humans, animals and plants increases, we have the exciting prospects of developing new ways to conquer or prevent disease, increase the productivity and hardiness of crops and improve the quality of life for the disabled. This growing body of knowledge and its ethical application should not be a cause for suspicion or fear.

The Origin of Life

For all that we have learned about the nature of life, one question remains unanswered: How did life come into existence? Of course, the simplistic answer is that God created life—even if it may have been only a single life form from which all others evolved. For a scientist, this answer is not satisfactory. After all, if God created the universe over billions of years in accordance with natural laws, isn't it also likely that he used similar natural means to create life? And if so, isn't it part of God's plan that we should 'think his thoughts after him' and discover the secret of the creation of life?

The as-yet unknown natural process (or processes), by which life arose from non-living matter, such as simple organic compounds, is referred to as ***abiogenesis***.

There are three known domains of life: *Bacteria*, *Archaea* and *Eukariota*. They are all believed to have from an unknown common ancestor. The Eukariota - which include all plants, animals, fungi, algae and protozoa - have clearly changed a great deal from the original ancestor. Bacteria and Archaea tend to be much smaller than eukariote cells and are much simpler in form and function. Both exploit a wider range of energy sources than can eukariote cells. Archaea are in many respects similar to bacteria but exhibit differences in their biochemistry from other forms of life and can flourish in extreme environments that would be fatal to anything else.

The most basic form of life is considered to be a simple single cell form, so it is reasonable to expect that the first life form was very similar to single cell bacteria or archaea. It is probable that it would have been adapted to one particular environment and energy source.

The age of the Earth is about 4.54 billion years; the earliest undisputed evidence of life on Earth dates from at least 3.5 billion years ago, and possibly as early as between 3.6 and 4.0 billion years ago, after the Earth's crust started to solidify. In May 2017 scientists found possible evidence of early life on land in 3.48-billion-year-old geyserite and other related mineral deposits (often found around hot springs and geysers) in the Pilbara region of Western Australia. Other discoveries suggest that life may have appeared on Earth even earlier - fossilised microorganisms within hydrothermal-vent precipitates dated from 3.77 to 4.28 billion years old have been found in rocks in Quebec, Canada, suggesting that life could have started soon after ocean formation 4.4 billion years ago

All currently known organisms rely on DNA to create proteins which in turn control cellular processes; but proteins are large molecules consisting of thousands of atoms. As such, they are not likely to have been around for the first organisms to use. Accordingly, most scientists involved in the field are inclined to look for simpler processes that could give rise to life from within primordial chemical environments of the 'young' Earth.

The first experiments designed to investigate the idea of life arising under these conditions came from the famous 1953 experiment by graduate student Stanley Miller under his supervisor Harold Urey. They tried to recreate the conditions on early lifeless Earth, with an atmosphere full of simple gases and laced with lightning storms. They filled a flask with water, methane, ammonia and hydrogen and sent sparks of electricity through them. When Miller analysed samples from the flask, he found five different amino acids—the building blocks of proteins and essential components of life.

Over the next decade, Miller repeated his original experiment with several twists. He injected hot steam into the electrified chamber to simulate an erupting volcano, another mainstay of our primordial planet. He also varied the gases by adding hydrogen sulphide and carbon dioxide to ammonia and methane.

If amino acids could come together out of raw ingredients, then bigger, more complex molecules could presumably form given enough time. Biologists have devised various scenarios in which this assemblage takes place in tidal pools, near underwater volcanic vents, on the surface of clay sediments, or even in outer space.

Miller's experiments showed that it was possible to create amino acids, but totally failed to create more complex molecules such as the nucleic acids.

The main problem with the study is that Miller was almost certainly wrong about the conditions on early Earth. By analysing ancient rocks, scientists have since found that Earth was never rich in methane, hydrogen sulphide or hydrogen itself. A more realistic mixture, mostly carbon dioxide and nitrogen, with just trace amounts of other gases, is unlikely to produce any significant amino acids.

The idea that the first life could have been based on DNA faces the problem that proteins are needed to replicate DNA, but DNA is necessary to instruct the building of proteins. Some researchers think that RNA (ribonucleic acid, a simpler cousin of DNA) may have been the first complex molecule on which life was based. RNA carries genetic information like DNA, but it can also direct chemical reactions as proteins do.

Recently, an undersea microbe, *Methanosarcina acetivorans*, has been discovered which eats carbon monoxide and expels methane and acetic acid. Biologist James Ferry and geochemist Christopher House from Penn State University found that this primitive organism gets its energy from a reaction between the acetic acid and the mineral iron sulphide. Other energy-harnessing processes require dozens of proteins, but this acetate-based reaction runs with the help of only two simple proteins. They suggest that this geochemical cycle could have been what the first organisms used to power their growth. This is significant because there is a growing acceptance among scientists in the field that deep-sea vents are a more likely location for life to have originated. These rocky 'chimneys' spew out superheated water and hydrogen-rich gases and are rich in minerals that are important for the healthy maintenance of life.

In a fanciful alternative explanation for the existence of life, some scientists suggest that life may have begun shortly after the Big Bang during a habitable epoch when the universe was only 10 to 17 million years old. This '*panspermia*' hypothesis proposes that microscopic life was distributed throughout the universe and that the early Earth could have received some of this life via space debris. This hypothesis does not explain how life originated; it simply transfers the point of origin to outside of this world. Apart from its lack of a testable process of life creation, it has the additional challenges of explaining how enough of the elements necessary for life could have been created through only a few generations of stars and how that 'life' could have endured billions of years of near absolute zero temperatures and intense radiation during its journey or how it survived during its likely fiery entry through Earth's atmosphere.

While the 'how?' and 'where?' of the origin of life are still mysteries, prevailing scientific thought is that transition from non-living to living entities was not a single event but a gradual process progressing through molecular self-replication, self-assembly, autocatalysis and the emergence of cell membranes. At this stage in our growth of scientific knowledge, we can conjecture about the necessary requirements for the creation of life but have no real idea about how such complex chemistry could have come about. Perhaps we never will.

What Does the Existence of Life Tell Us About God?

Life exists. We exist. Was it by accident?

Readers who are familiar with the Old and New Testaments are aware that there are many stories about God intervening in history to act outside of what we understand to be natural laws—such as the parting of the Red Sea (Exodus 14:10-31) or the calming of the storm on the Sea of Galilee (Mark 4:35-41). Indeed, many of you will know of or even have experienced for yourselves miraculous healings or other instances of divine intervention in response to prayer. So it is not unreasonable to think that the creation of life, at least in its simplest form, could have been as a result of divine action instead of the random conjunction of suitable amino acids in an environment conducive to the creation of 'life'. However, I would warn my readers of the dangers of depending on this explanation as evidence of a God.

In the past, many people have pointed to 'gaps' in our scientific knowledge, using them to argue the need for a God. Of course, as our knowledge of the world and universe grows, these gaps have been reducing in number and significance. Is it any wonder then that many people object to 'having their arms twisted' to believe in a God as the solution to any unexplained phenomenon. Remember Laplace's statement 'Sire, I have no need of that hypothesis.'

Many years ago now, the New South Wales Teachers Christian Fellowship and the Sydney University-based Research Scientists Christian Fellowship joined together to conduct a week-end seminar for high school science teachers. During one of the sessions that I chaired, an elderly lady science teacher confessed that she was afraid of believing in the theory of evolution for fear, as she expressed it, 'that it might destroy my faith'. My response was that we should never be afraid of the truth—if mankind can be demonstrated to have evolved from simpler forms of life then that truth must form part of our understanding of God and his purposes. Our responsibility is to look at the whole of God's revelation of himself and his creation, both through his inspired word and through science, in order to deepen our knowledge of him and of the powers that he has made available to us in serving him and each other.

So, if you are still depending on a literal interpretation of the *Genesis* account and we eventually discover the mechanism and conditions for the production of life and perhaps even succeed in creating 'life' itself, what is going to happen to your belief in God?

If we are really meant to 'think God's thoughts after him', we need to dispense with this 'God of the Gaps' mentality and pursue with vigour every line of investigation into the wonders of his creation. It is time for us to stop believing in God because we need him to have created the universe or because we need him to have created life. As we have seen in Chapter 3, creation tells us nothing about the Creator except the Creator's great power and existence independently of this universe. It is time we started believing in God because he has reached out to us.

7

THE ASCENDANCY OF MAN

What is the Difference?

In Genesis 1 we read:

> [24] And God said, "Let the earth bring forth living creatures according to their kinds--livestock and creeping things and beasts of the earth according to their kinds." And it was so.
> [25] And God made the beasts of the earth according to their kinds and the livestock according to their kinds, and everything that creeps on the ground according to its kind. And God saw that it was good.
> [26] Then God said, "Let us make man in our image, after our likeness. And let them have dominion over the fish of the sea and over the birds of the heavens and over the livestock and over all the earth and over every creeping thing that creeps on the earth."
> [27] So God created man in his own image, in the image of God he created him; male and female he created them.

Here we have a separation between animal life in its huge variety and mankind. Now there is a world of difference between an elephant and a mouse, between a lion and a dung beetle, between an eagle and a salmon, and yet these are lumped together as part of a common creation. On the other hand, humankind is accorded its own special creation. Why?

The explanation offered here is that humankind is made 'in the image (Hebrew: *tselem*) of God' and with a special purpose: 'so that they may rule over (Hebrew: *râdâh*)' all other forms of life'.

Within this context, the Hebrew word *tselem***,** (tseh'-lem) has the meaning of 'resemblance' or even 'a representative figure'.

In the same context, *râdâh* (raw-daw) literally means 'to tread down' or 'to crumble' (perhaps a very fitting description considering the immense damage that we have done in wiping out whole species), and is used to mean 'subjugate', 'dominate' or 'prevail against'.

So, in the light of science, what are we to make of this? If all life, including man, evolved from a single original life form, what is so special about us that allowed us to predominate?

The 'Missing Link'

Despite popular mythology, man did not descend from the apes. Rather, evolution maintains that man and the apes descended from a common ancestor. Almost from the beginning of the argument over creation versus evolution it was popular to talk about a 'missing link' between apes and humankind, as if somewhere in prehistory there was single creature that would mark the time when apes and humans diverged.

At the beginning of my university education I was a '*creationist*'; *Genesis* told me that God created the heavens and the earth in six days, so that was what I believed. I maintained (as many still do) that science has never identified a 'missing link' that showed human descent from ape-like creatures. A biologist friend that I worked with at the Australian Atomic Energy

Commission patiently explained to me that there could never be a single missing link, but that there had to be hundreds, perhaps thousands, of generations of 'missing links'.

In the six decades since then there has been an 'explosion' in knowledge of the process and timescale of evolution. All around the world, scientists have been uncovering and extending the fossil record of life on Earth and, in the process, validating the theory that more complex life forms appear after and probably developed from simpler life forms.

In addition, a massive amount of work is going on in mapping the genomes (the DNA coding possessed by an organism that determines the kind of life form that it will become) of every kind of life form from bacteria to humans. This work has been critical, not only for understanding where the differences lie between species and between individuals within a specie but also at what point in geological and evolutionary history divergences between various life forms occurred.

Today, we can with some confidence chart a broad timeline of the development of life on Earth. You can easily find this on the Internet, so I will not devote much time to this subject here.

As mentioned in Chapter 6, evidence from the fossil record for the earliest life forms suggests that life began at least 3.48 billion years ago. At that time Earth was inhospitable to life as we know it; although liquid water had been present for over 100 million years, the planet was very volcanic and the atmosphere was totally lacking in oxygen. It was not until about 3 billion years ago that photosynthesizing cyanobacteria evolved; these produced oxygen as a waste product, which allowed the oxygen content of the atmosphere slowly to build up.

Fast forward now to ten million years ago ... Our closest living relative to man is the chimpanzee. About ten million years ago, the ancestors of man and chimpanzee began to differentiate and they were clearly differentiated by four million years ago.

By three million years ago, our pre-human ancestors had become fully bipedal (walking on two legs), had lost much of their body hair and were making simple tools. Between two and a half and two million years ago, *Homo habilis* appeared—possibly the first of our forebears to master stone tool technology. By two million years ago, some of these early proto-humans had spread out from Africa to as far away as China.

Around 800,000 years ago, our ancestors began to diverge from Neanderthals and in the next 100,000—200,000 years or so, Neanderthals and a third human group, the Denisovans, began to diverge from each other. The earliest evidence of the appearance of the first 'modern' humans, *Homo sapiens*, dates from around 200,000 years ago, although some archaeological evidence suggests that they may have existed as early as 300,000 years ago. Neanderthals and Denisovans have been extinct for at least 40,000 years, but there is evidence of limited interbreeding between Homo sapiens, Neanderthals and Denisovans prior to that time; about twenty percent of modern human groupings have Neanderthal and/or Denisovan DNA in their genetic make-up.

What Distinguishes Mankind from Other Animals?

If we are to believe that humankind has something that marks us out as being 'in the image of God', something that is unique to mankind, what is it?

At one time scientists thought that the manufacture and use of tools was unique to man. We now know that simple tools were in use by our predecessors for at least three million years. Neanderthals and Denisovans used sophisticated stone tools alongside Homo sapiens, but we also know that a number of modern non-human species utilise simple materials as tools; for instance, chimpanzees use grass stalks to catch termites or stones to crack nuts, some birds of prey have learned to drop tortoises and other hard-shelled creatures from great heights onto rocks to crack their shells and a type of Australian kite is known to pluck burning sticks from a recent fire and to use them to start fresh fires elsewhere in order to flush out the insects that comprise their diet.

There is no doubt that the level of intelligence and the complexity of brain structure are greatest in humans, but this difference is largely a matter of degree. The extent to which humans have excelled in the development of their brains is almost certainly due to the variety of the environmental challenges that mankind and their hominid forebears have encountered and to which they adapted as they spread out across the globe. Humans have become the ultimate survivors not just because of their ability to adapt to an environment, but also because of their ability to exploit and modify that environment.

But large numbers of modern animal species have complex brains and exhibit great intelligence. The Internet is full of videos of animals and birds displaying amazing levels of skill in problem solving that require analysis of the problem and the logical identification and ordering of the necessary actions to achieve desired outcomes. So it is not the lack of these skills that limit the brain development of most animal life—it is the environments that they are tied to (even migrating species); there is insufficient pressure on them to change.

Again, even some traits that we like to think of as 'truly human' can be found in the animal world - traits such as compassion. Again, the Internet is full of videos of individuals of one specie rescuing or protecting individuals of other species, whether of an ape rescuing a bird from drowning or a lioness protecting a baby gazelle from a pride of hungry lions.

There is a body of scientists that considers that the only unique quality in mankind is that we, of all life forms, are aware of our place in this world and the universe; all other 'intelligent' life forms, as far as we know, are aware only of their immediate environment and their concern is solely with their own survival and the drive to continue their species.

On the other hand, Christian author, broadcaster and lay theologian C S Lewis maintained that the existence of moral law (an innate knowledge of the difference between right and wrong) within mankind is unique amongst all animal life and is clear evidence of a God that cares about us as individuals. This argument was very influential in convincing Doctor (now Professor) Francis Collins, leader of the Human Genome Project in the first decade of this century, of the existence of God, turning him from an atheist to an active Christian layman. He says that he rejected the idea that the moral law is a consequence of evolution because moral law appears often to conflict destructively with what may be best for our individual or group survival. Certainly it seems that a sense of right and wrong is universal amongst all peoples and is an essential component of shared understanding in any society's attempts to live together. However, it must be pointed out that this could be a learned trait that develops as a child matures within a family or social group.

Something that almost all researchers would agree on is that only humans seem to have an inbuilt belief that something or someone greater than us exists with powers that more or less control our lives. In every continent and in virtually every grouping of peoples, there is a belief in gods or spirits and that we need to please or placate it or them in order to obtain favour on our own behalf. This belief is losing ground as we better understand the nature of the world in which we live; this is a good indication that the universality of this phenomenon is not a product of evolution.

While these factors seem to be unique to humankind and clearly are not 'outcomes driven' by evolution, there is good reason to regard them not so much as an innate faculty but as a 'learned' one, much as the development of the use of simple tools, but requiring a very much higher level of development of another faculty—intelligent communication.

What Do We Mean by 'Intelligent Communication'?

'Intelligent communication' refers to the intentional passing of meaningful information by one individual or group to another individual or group.

All social animals—even insects—have developed means to communicate with others of their kind. For many, this communication is limited to physical display to get their message across (for instance: bees 'dancing' to indicate to other bees the direction and distance of a source of pollen that the bee has located; some varieties of birds that 'dance' with their partners to establish and maintain their relationship). Typically, the young learn by observing the behaviour of their elders. Some life forms communicate by 'smell' to advertise sexual receptivity. Ecologist Suzanne Simard discovered that even trees communicate their needs and send each other nutrients via a network of latticed fungi buried in the soil.

However, the most common form of animal communication uses sound because this is effective over distances, even where the individuals can't see one another—sounds that mean 'I am here' or 'I am looking for a sexual partner' or 'Watch out! There is a predator about'. Like physical displays, their purpose is to inform the hearers and can be used to reassure, warn or attract. Some animals and birds can differentiate warnings to indicate the nature of a threat—does it come from the sky or is it stealthily making its way through long grass?

Perhaps mankind's greatest evolutionary advantage is that of speech. A necessary requirement for speech is the presence of the hyoid bone (the only bone in our bodies that is not attached to any other bone). This development occurred sometime after the divergence of *Homo* from the precursors of the great apes (between two million and 800,000 years ago). Hyoid bones have been found in association with Neanderthal fossils, suggesting that Neanderthals and Denisovans shared our own ability to speak, but to what degree is not clear.

We can only guess how language developed, but it is reasonable to assume that it began with similar meanings to other forms of sound communication such as expressions of warning, greeting or reassurance. As the ability to make and control sounds improved, it is probable that language began to introduce names for various types of animals and objects encountered in a group's movements, much as Adam is said in Genesis 2 to have given names to all of the creatures that he encountered. Around the same time, words would have been developed to describe simple actions (walk, eat, drink, etc.). Inevitably, adjectives were needed to better describe an animal or object (big, little, cold, hot, wet, dry) and actions (verbs) would be modified by adverbs (softly, slowly, quickly). All of this would have required a basic syntax (rules about how words are organised) in order for the meaning to be clear. With these basic concepts in place, it would not have been too big a leap for new words to be introduced to express emotions, share thoughts and exchange ideas.

It was language that allowed humans to begin to contemplate their place in the universe and allowed the development of myths and legends to give them a sense of belonging. It was language that allowed the deeds of individuals to be celebrated and remembered. And it was these myths and stories of past glories that encouraged individuals and groups to aspire to their own achievements.

Initially, language was only spoken. Before the invention of writing, the only way to preserve a group's history and creation myths was for a group to appoint one or more individuals who were entrusted with the responsibility of memorising those stories, of regularly reciting them to the group and teaching them to selected persons of the next generation who were to inherit the role after them.

The oldest known drawings on rock walls date from about 40,000 years ago, but the oldest known writing found to date (from Greece, China and Romania) all date from the 6th Millennium BC, are generally symbolic and may not display any significant textual meaning. Most early writing tended to be *logographic* (a symbol representing a whole word or even a complete phrase, much like Chinese script). These hieroglyphic symbols appeared in both Egypt and Mesopotamia. The earliest known written records of Egyptian myths are in hieroglyphs and date from around 4000 BC.

Much of early writing and numbering probably arose out of the need to keep accounts in relation to trade over long distances. The original Mesopotamian writing system was derived around 3200 BC and was made as indentations in soft clay (one mark for each item) associated with pictograms of the things being counted. Where the numbers involved were large, this was clearly impractical and, by 3000 BC, they were using a simple set of triangle-based symbols to represent larger groupings such hundreds and thousands. By the 29th Century BC symbols were being included that represented phonetic elements and around 2700 BC, Mesopotamian cuneiform began to represent syllables of spoken Sumerian.

With the development of written languages, mankind has had the ability to share information at great distances and across long periods of time, limited only by the quality of the materials on and with which it is recorded. We are no longer limited by the frailty of human memory. There is no doubt that it has been our ability to record, preserve and share our thoughts and experiences in words and pictures (and now digitally) that has led to the explosive progress of mankind over the last two hundred generations in understanding ourselves, our world, our universe and our God.

8

FIRST CONTACT

So far in our considerations we have seen that there is a huge difference between the simplicity of the *Genesis* accounts of the creation of Earth, the universe and life itself and the incredible complexity of the scientific reality. I have maintained that the *Genesis* stories were geared for the minds of primitive man with the primary concern being to emphasise the power of God the Creator and God's intention that humankind should work cooperatively with God to understand and gain domination over the whole world.

Now we need to look more deeply into the progressive revelation of God's own nature and plan to that same primitive level of understanding as they unfold in Chapters 2 and 3 of *Genesis*.

Readers who are familiar with the Bible will be aware of the many times and many ways that God has revealed himself to and interacted with individuals, groups and even huge crowds. So it is only logical to recognise that at some point in time there was a first instance of God revealing himself to humankind. The story of Adam and Eve in Chapter 2 of *Genesis* claims to be an account of that first contact.

We should probably look first at the human characters involved. The Hebrew word *adam* literally means 'man' or 'human' and in most English translations of the Bible is rather indiscriminately translated as 'Adam' or 'the man' in various places in the *Genesis* account. The Hebrew text itself does not distinguish any of the occurrences as intended to be a given name. Eve, on the other hand, is a name specifically given to his female partner by the man (Gen. 3:20). 'Eve' is an anglicised version of *chavvah*, meaning 'life' or 'living' because she is expected to be the 'mother of all living'. 'Adam' becomes a proper name only when the Hebrew word *adam* was transcribed literally into Greek.

What can we say about the story of 'Adam' and Eve? Firstly, it can only have had one of two origins: (1) that it is a myth—a story wholly made up to entertain or provide some kind of explanation of a tribe's origins; or (2) that is based on true events, the central details of which must have originated with Adam himself. If a myth, it could well have originated quite late in the tribe's history, maybe even subsequent to the invention of writing, but probably based on some details that had been passed down from some indeterminate time in the past; if based in fact, it may have been passed down solely by word of mouth for an unknown number of generations before ever being written down. In either case, it could have suffered any number of changes to enhance the story for its listeners, but it is possible that the core facts would have been preserved.

While many theologians consider the account to be mythical, many others accept the story to be wholly or partially factual, believing that God is well able to preserve his own revealed truth. With the caveats outlined below, I place myself firmly in the second group because the lessons taught in Chapters 2 and 3 of *Genesis* are squarely in keeping with the nature of God and his plan for mankind outlined consistently throughout the Old and New Testaments.

However, if we are to give any credence to the historical existance of Adam and Eve and the events described in chapters 2 and 3 of *Genesis*, then we should expect that our current scientific knowledge could help us to position them within a specific geographic region and historical age. If this can be done, then it might be possible to better identify and understand both their historical symbolic significance.

The 'Garden of Eden'

Genesis states that God established a garden within a region referred to as *Eden* and it was there that the first contact with mankind occurred. The only clues to its location was that out of it flowed a river that then became the headwaters of four rivers—the *Pishon*, the *Gihon*, the *Tigris* and the *Euphrates*. The first two of these rivers are unknown today and the regions that they are said to have watered, like *Eden* itself, are not identifiable from their names or descriptions. If the story is based on real people in a real area, it is almost certain that the location lay in or near the Armenian Highlands of Turkey where both the Euphrates and Tigris rivers originate. This is supported by the fact that gold is mined in the general area and the river *Pishon* is stated to have flowed through a gold-bearing area.

A major problem faced by archaeologists and geologists in pinpointing the location is the time that has elapsed between then and now—a period of somewhere between eight and thirteen thousand years. This time span is certainly long enough for significant geological events (especially in such an active tectonic region) and climate changes to have brought about immense changes in topography of the land and the nature of vegetation. Around the world and even within the Mediterranean and Middle Eastern regions, these types of changes have been demonstrated to have occurred in much shorter time frames.

Placing the events of Chapters 2 and 3 of *Genesis* within world pre-history is only slightly easier, although still based on 'best guesses'. The last ice age began to recede around 11700 BC and was completely finished around 11000

BC; it is unlikely that a naked man was able to live in that region prior to that time. Archaeological research in the 'Fertile Crescent' and Mesopotamia indicates that from around 11000 BC the whole Middle Eastern area was widely but sparsely populated, mostly by small familial groups that were essentially nomadic hunter-gatherers. The period from 10000 BC to 4500 BC is popularly known as the *Neolithic* (New Stone Age).

During the years 11000 - 7000 BC, the practices of agriculture and animal husbandry in the area began and gradually expanded to become the predominant way of life, resulting in the growth of permanent settlements with simple permanent structures.

The setting of the early chapters of *Genesis* seems to fit well into the earliest part of the period between 11000 BC and 9000 BC. It is easy to picture a lone explorer wandering into the Armenian Highlands and finding a verdant wilderness rich in edible plants and fruit-bearing trees. It is in that context that I believe that God could have initiated his first human contact.

We need to recognise that worthwhile contact could only have occurred after *Homo sapiens* was already fairly well established, had a well-developed language and had some basic social conventions to be observed when groups met. We have no solid evidence of any formal religious practices at this early date except in relation to burial of the dead; the earliest clearly identifiable temple in Mesopotamia was not built until around 4500 BC and the earliest written records of religious practices in Egypt date from 3400 BC.

However, it is likely that the concepts of a god (or gods) and a 'spirit world' were already familiar subjects for conjecture amongst the peoples of the time. This is borne out by the fact that a belief in gods and spirits is universal in native populations around the world. The Australian indigenous peoples have been virtually isolated on the Australian continent for about 60,000 years yet have a rich spirituality and mythology.

Recent excavations at Gobekli Tepe in south-eastern Turkey have found large circular stone pavements and pillars that could date from as early as 9500 BC and the area appears to have been a meeting place where hunter-gatherer peoples met. Some have suggested that this might even have been a very early temple. Around 10700–9400 BC, a settlement was established in Tell Qaramel, a few miles north of Aleppo. The settlement included what appear to be two temples dating to 9650 BC. In light of this, it does appear that regular religious beliefs and practices are more likely to have developed in the Turkey-Northern Syrian region that within Mesopotamia.

We can only speculate on human ideas of the time as to how to communicate with, placate or gain favour with the gods.

Encountering the Divine

In what way could an encounter between God and Adam have taken place? In later times, men experienced God in many ways—dreams, visions, angelic visitations, voices from heaven as well as God himself taking on the appearance of a human. *Genesis* tells us that this first series of meetings (which we are told took place over an extended time period that could have been anything from days to months) was certainly one in which God appeared in physical (probably human-like) form. Genesis 3:8-9 says:

> *[8] And they heard the sound of the LORD God walking in the garden in the cool of the day, and the man and his wife hid themselves from the presence of the LORD God among the trees of the garden.*
> *[9] But the LORD God called to the man and said to him, "Where are you?"*

But if the man encountered God in simply human form, what convinced him that the Lord God really was God? And what did the man understand by the concept of a god in the first place?

There seems little doubt that God asserted his ownership of the area and that this was accepted by the man without argument; there was no confrontation as could normally be expected over a piece of prime real estate. Instead, the man is given free run of the place and responsibility for oversight and care of the 'garden'.

It seems clear that during this period the man gained a mate. Did the man bring with him a female companion or did God prompt a young woman to also enter this area or did the man leave for a time in order to seek a mate amongst other nomadic groups? We are given no understanding of how Eve arrived on the scene (except that it was apparently at God's instigation, but it was certainly God's intention that this man and woman were to become the basis for a familial group who were to be the means by which he would reveal himself to all mankind.

This period could justifiably be regarded, if not literally as the 'beginning' of humankind, then at least the beginning of humankind's special relationship with God. In this sense then, Eve could certainly be regarded as the mother of all who were to be truly 'alive' in terms of a privileged spiritual relationship with the Creator.

While we have no external corroborating evidence as to how historically accurate the *Genesis* account of this first contact might be in all of its details, the clear purpose was to help us to understand that it is God who wants to enter into a close relationship with his creation and has initiated this on his own terms. On this basis, I believe that our main concern should be with what the Creator was revealing about himself to the couple rather than on details about when and how this contact was first established.

9

CHOICE AND FREE WILL

The Two Trees

Genesis 2:9 tells us that God placed two trees in the centre of the garden—*the tree of life* and *the tree of the knowledge of good and evil*. In verses 16 and 17 of the same chapter, God tells Adam that he can eat from any of the trees in the garden except *the tree of the knowledge of good and evil*. He is also warned that 'on the day (Hebrew *yowm*) you eat it you will surely die.' [most English versions]. The New International Version translates the passage as 'when you eat it …', choosing to interpret *yowm* to mean a more indeterminate period of time. Perhaps this was done to avoid the problem of Adam surviving for a considerable number years after his sin. This interpretation is unnecessary, as we shall see.

Why did God place a tree in the garden that he specifically didn't want Adam to eat from? How could God be so provocative as to confront Adam with something so dangerous that to eat of it would result in immediate sentence of 'death'? What kind of a God deliberately risked this new relationship with his creation when it had hardly begun?

Regardless of the historical accuracy of the story as it has been passed down to us, there is no doubt that the story is a graphic example of the first lesson that humankind needed to learn: It is the Creator God who initiated and continues to offer the opportunity for humans to enter into a close personal relationship with him. And it is God who has the right to make that offer conditional. Throughout the entire history of the Jews and the early church, God made it clear that he wants that relationship to be based on our own willing acceptance of and submission to his authority. But free will implies that a choice has to be made; where there is no possibility of choice there can be no free will. So God provided Adam with a choice.

The choice is presented in terms of the two trees. From a botanical point of view it is not clear whether these two trees were unique in their character from any other trees found in nature or even just from any other trees in the garden. However, there is no doubting that these trees were unique in their immense symbolic significance: Choosing to take fruit from the *tree of life* (as we are told in Genesis 3:22) would result in them receiving eternal life; choosing to eat of *the tree of the knowledge of good and evil* would allow them to have a god-like understanding of good and evil that would allow them to live independently of God and to be masters of their own destiny. God's offer is being made clear in terms that a primitive man could understand.

The 'Fall' of Man

A new character enters the story at this point to make sure that Adam and Eve don't fail to grasp the full meaning of the choice they have to make. Existing ancient versions of *Genesis* identify this new character as *the serpent*. Later references throughout the New Testament make it clear that this was Satan, the Devil, the Accuser, the Father of Lies. It is not clear whether Satan took the form of a serpent or, perhaps more likely, possessing some serpent-like characteristics. It seems improbable that even the most primitive of men would happily converse with a snake and equally improbable that a snake could actually speak (although in Chapter 22 of the book of Numbers, God allows Balaam's donkey to speak). It may be possible that, over the generations through which this story passed, changes in words or their meanings may have occurred, but this seems unlikely given that Satan is referred to as 'that ancient serpent' in Revelation 12:9 and 20:2.

Regardless of the form in which Satan appeared, the question needs to be asked: Why did God allow the serpent into the garden? If it hadn't been for the serpent, perhaps Adam and Eve would never have eaten the forbidden fruit. Certainly, up to the time that the serpent appears, Adam and Eve seemed quite happy to obey the Lord God's command.

It is precisely that last point that made the presence of the serpent necessary; it was an essential part of God's plan that man should be confronted with the full implications of the choice to be made (although the serpent ignores the significance of *the tree of life* and concentrates on the benefits and desirability of the other tree).

Satan, 'the father of lies', tells the truth about *the tree of the knowledge of good and evil*, but, while it may have been in accordance with God's purpose, it is certainly to the serpent's own advantage. He certainly lies in denying the truth of God's warning that they would die if they ate from it. (It is interesting that when Satan was tempting Jesus in the wilderness he also spoke the truth when he claimed to have the right to give to Jesus all the authority and splendour of all the kingdoms of the world. If that had been a lie, Jesus would have known the truth and Satan's offer would not have been a real temptation.)

Most readers will have all heard of the *seven deadly sins*, but no such classification is made in the Bible. Instead, in 1 John 2:16, sin is classified under three headings - the lust of the flesh, the lust of the eyes and the pride of life. Each of these classes of sin were represented in Satan's temptation of Jesus and all three aspects were active in Eve's decision to disobey God's command—*'So when the woman saw that the tree was good for food, and that it was a delight to the eyes, and that the tree was to be desired to make one wise, she took of its fruit and ate, and she also gave some to her husband who was with her, and he ate.'* (Gen 3:6)

The story goes on to say that the eyes of both of them were opened and they realised that they were naked. They were ashamed and tried to clothe their nakedness. Note that it was not God who was offended by their naked bodies; they had been associating with God for some time and God had not declared them naked or taken steps to clothe them. Nakedness was not an issue in God's sight. But Adam and Eve were now making their own judgements as to what is right and wrong and what concerned them most was external appearances.

But now they found that their true nakedness was not external—God could see into their thoughts and motivations. They had made their decision. And now, like naughty children (and like us all), each tried to shift responsibility—Adam: 'It was the woman'; Eve: 'It was the serpent.'

Judgment

Among those Christians who believe in the literal six days of creation many also believe that Adam's sin brought death into the world and that if Adam had not sinned there would be no natural death of any form of life. This is foolish on three counts:

- Scientists are fully aware of the multiplicity of generations of life that had existed prior to Adam's time. In fact, many species appeared and then became extinct before humans ever appeared on Earth (dinosaurs being a comparatively recent example);
- If death came only with Adam's sin, then if Adam had never sinned, the world would quite quickly have become totally unin-

habitable; the sheer volume of insects, which produce prolific numbers of progeny in timespans of days, weeks or months, would have resulted in the available liveable space on the ground and the breathable air around us being taken up by insects to the extent that normal mammalian life would be untenable. On this basis, God's plan would have been thwarted; and

- God states in Genesis 3:22 that if Adam was to take of the fruit of the tree of life even after being judged, he would still have received eternal life, thus implying that until he tasted that fruit he was mortal.

But, if physical death did not appear in the world as a result of Adam's sin, what are we to make of God's statement that 'on the day' that Adam disobeyed, he would die? Clearly that did not happen. So, was God lying? Did he change his mind? Or are we to look for a deeper interpretation?

The apostle Paul makes it quite clear (in Ephesians 2:1, 5 and in Colossians 2:13) that, in our sinfulness, we are 'dead'. By this he is saying that our sins separate us from God, the source of life, and as a result we are spiritually dead. And this is exactly what God meant in his warning to Adam—choosing to be independent of God, to be their own best judge of what was right for them, meant that Adam and Eve were shut out from the close personal fellowship with God that they had shared. On that day, the special relationship was broken and sin and spiritual death entered the world.

It is necessary at this stage to understand that Paul, in many of his letters to the churches, distinguishes between *sin* as a principle or motivating power and *sins*, which are actions arising resulting from that principle. Adam and Eve *sinned* (action) in deciding to eat of the forbidden fruit, but what they were taking into their lives was the sin principle—that attitude that chooses the right to be independent of God, to make one's own decisions based on one's own judgements and desires.

Romans 5:12 makes it clear that sin and (spiritual) death entered into the world through Adam and Eve. Romans 5:13 tells us that sin (the principle) was in the world but it is not taken into account when there is no law. When Adam and Eve disobeyed God's only command to them, that sin was taken into account. No-one else at or prior to that time had sinned because, for those others, there was no revelation of God's standards and no direct relationship with God. Whether or not the story is literally true in all its details, its true significance is as an explanation of just what sin is and what its consequences are.

But how could Adam's sin inflict spiritual death on all men? Are sinfulness and spiritual death things that we inherit? And if it was, how could it be inherited by someone who didn't actually descend from Adam's line?

Perhaps this is a good time to revisit the subject of a moral law mentioned in Chapter 7 of this book. In the second chapter of Romans, Paul writes:

> *[12] For all who have sinned without the law will also perish without the law, and all who have sinned under the law will be judged by the law.*
> *[13] For it is not the hearers of the law who are righteous before God, but the doers of the law who will be justified.*
> *[14] For when Gentiles, who do not have the law, by nature do what the law requires, they are a law to themselves, even though they do not have the law.*
> *[15] They show that the work of the law is written on their hearts, while their conscience also bears witness, and their conflicting thoughts accuse or even excuse them*

> *16 on that day when, according to my gospel, God judges the secrets of men by Christ Jesus.*

Paul seems here to accept the idea of a moral law that can influence a man's actions in the absence of any written code of behaviour—people living in accordance with or contrary to that inner sense of right and wrong will receive the same judgement as those who were aware of the published law and chose to obey or break it. If this is so, then clearly Paul in Chapter 5, verse 12 of the same letter cannot be suggesting that we all are all spiritually dead because of Adam's sin. Indeed, Paul states that all are spiritually dead because 'all have sinned.' It is our own efforts to live independently of God that constitutes the sin of which we are all guilty and which has produced the same consequences—separation from a personal relationship with him.

The judgments decreed by God on Adam and Eve in Genesis 3:16-19 are not additions, nor are they a merciful alternative to God's original declaration that they would die. God is simply pointing out the implications of their desire to be independent. God is now no longer obliged to provide the wherewithal to sustain their lives or establish their family line. They are now their own responsibility, with all of the blood, sweat and tears that this will entail.

Suggestions that Satan took on the form of a serpent are difficult to take literally, in part for the reasons I mentioned earlier. However, God's initial judgement of Satan in the garden (Gen. 3:14-15) makes a literal interpretation even more difficult; did the serpent have legs before this time? (If so, two, four, six or more—perhaps like a caterpillar or millipede?) Science is not aware of any legged snake species, although we do know that there are legless lizards. It would be more sensible to see this judgement passed on the serpent as being entirely symbolic. Satan is degraded and, in his ongoing enmity against mankind, one of Adam's descendants will be wounded by Satan but that descendant will deal Satan a mortal blow. Thus, in the very earliest writings that constitute the Bible, the coming of Jesus and his victory over sin is already prophesied.

Genesis 3:21 tells us that God made garments of skin to clothe Adam and Eve. In ancient Mesopotamia during the *Neolithic* period (10000-7000 BC) it could be possible that clothing was not widely known or used because of the mild climate. However, scientific research indicates that clothing made from skins had been used perhaps as far back as 170,000 years ago in Northern Europe, Russia and China (and would have been essential during the last great ice age). Bone needles have been found in all of these areas dating from 15,000 to 40,000 years ago. In fact, one needle found in the Denisova cave in Siberia dated to around 50,000 years ago. This cave was used continuously by Denisovans and Neanderthals between 50,000 and 30,000 years ago.

Cain and Abel

Chapter 4 of *Genesis* is the last part of the section referred to as *the account of the heavens and the earth when they were created* (Gen 2:4) and is primarily concerned with two lines of descent from Adam and Eve.

Cain and Abel are identified as the first two of Adam's sons. No information is provided about how old Adam was when either of them were born, but the third son was born when Adam 'had lived' 130 years (Gen. 5:3) and this was clearly after the death of Abel (Gen. 4:25). Despite doubts about the accuracy of the some of the recorded details about these two men, based as they are on spoken accounts handed down over many generations before ever being first written down, they do provide some interesting insights into the culture of the times and the ongoing relationship between God and Adam's line. The first of these are the occupations of the two brothers.

Genesis 4:2 tells us that Cain *worked the soil* (*'abad*—worked, served; *'adamah*—ground) while Abel *kept flocks* (*ra'ah*—kept, tended; *t'son*—sheep, goats). Archaeological research has established that domestication of sheep began in Mesopotamia as early as 11000 BC and was fully established by 9000 BC. Domestic planting and harvesting of grain crops in the Fertile Crescent was also well established by 9000 BC, although archaeological evidence indicates that organised farming of naturally-occurring vegetables and fruits in the area had been in existence from as early as 10000 BC. On this basis, we can reasonably assign Cain and Abel to a period no earlier than 10000-9000 BC.

Events leading to the killing of Abel by Cain are found in Genesis 4:3-7:

> *³ In the course of time Cain brought to the LORD an offering of the fruit of the ground,*
> *⁴ and Abel also brought of the firstborn of his flock and of their fat portions. And the LORD had regard for Abel and his offering,*
> *⁵ but for Cain and his offering he had no regard. So Cain was very angry, and his face fell.*
> *⁶ The LORD said to Cain, "Why are you angry, and why has your face fallen?*
> *⁷ If you do well, will you not be accepted? And if you do not do well, sin is crouching at the door. Its desire is contrary to you, but you must rule over it."*

If this is an accurate reflection of events, we must conclude that God was still available as some type of physical presence or manifestation for

Adam and Eve and their children after being cut off from the garden; this is evident from both Cain's conversations with God and the fact that both sons brought offerings to God at a time when there was no 'holy place' where they could expect to leave their offerings. Indeed, it is clear that they both had feedback on the acceptability of their gifts.

There have been disagreements between theologians as to why Abel's offering was acceptable to God while Cain's was not. Some have suggested that it may have been due to Abel's being a sacrifice of blood. I tend to believe that God would not have rejected a non-blood offering from someone who by their profession was a worker of the soil; certainly, there is no indication in the story that this might have been the case. I think the real significance of the discrimination lies in two words that describe the offerings.

We are told that Cain brought *an offering of the fruit of the ground*. The Hebrew word used for 'fruit' is *periy*, referring to the product of the plant in general. In the later books of Moses (*Exodus - Deuteronomy*), the Children of Israel are required to bring as an offering only the 'firstfruits' (Hebrew: *bikkuwr*) of their crops. We are told that Abel brought fat portions from some of the firstborn of his flock. Abel gave the best of the first whereas Cain's offering is more likely to have been an afterthought and perhaps even of left-overs. I believe the message intended for us as well as for Cain is clear—God wants and expects that he is recognised first and with our best.

The Judgment on Cain

Considering that God decrees the death penalty against those who break many of his laws laid down for the Children of Israel in Exodus, Leviticus, Numbers and Deuteronomy, it may seem surprising that Cain is not killed by God for the murder of his brother. Part of the apparent contradiction may be that in laying down the laws for the Hebrews God's first concern is to eliminate any possible on-going contamination of his people, and this is best done by swiftly removing the violent, debauched or rebellious elements.

In Cain's case, there is no such issue and God sentences Cain to a life-long exile (perhaps with the prospect of learning from his mistakes).

Cain's own attitude isn't in the least repentant and is more concerned with his own continued well-being. What is of genuine interest to us is that Cain recognises that there are other people spread around the area and is afraid of them. This is contrary to the suggestion that the total population of the world at that time consisted of Adam, Eve and Cain and confirms the modern scientific understanding of the region at that time.

Cain also acknowledges that he is being driven from the presence of God, again supporting the idea that God was known by Adam, Eve and Cain in physical form and/or a specific location.

We don't know how Cain was marked, but it clearly was in such a way that anyone who saw him, regardless of any knowledge they might have as to Cain's crime, would be instinctively afraid of or in awe of him so as to do him no harm. This might have been based on some superstitious belief amongst the people of the region where the 'mark' indicated that he was either special or dangerous; the fact that he was able to find a wife amongst these peoples (Gen. 4:17) perhaps suggests the former rather than the latter.

Cain, we are told, *went out from the Lord's presence* (again suggesting that God's presence was a physical one in a set location) *and lived in the Land of Nod* (Hebrew: *nowd*), *east of Eden*. The Hebrew word *nowd* means 'wandering' and could be a descriptive expression for a region where there were no regular settlements rather than a specific name for the area. It is in this area that Cain builds a 'city' and calls it after his son, Enoch. The Hebrew word translated as 'city' is *iyr* is actually used to describe any permanent or temporary settlement that is guarded and can be as simple as an encampment; so 'city' is probably an overly generous term. A number of characters mentioned in *Genesis* are said to have founded 'cities' named after themselves and it is more reasonable to think of them as one or more dwellings and referred to by the name of the builder, e.g. 'George's (place)'

Genesis only concerns itself with six generations of Cain's descendants and no indication is given of the age to which each of his male descendants lived; this family line is seen as having no significance in the ongoing story of God's dealings with mankind. However, two points can be made from this short genealogy:

God undertakes to ensure that anyone killing Cain '*will suffer vengeance seven times over*' (Gen. 4:15), presumably at the hands of God. In Gen. 4:23-24, Lamech *is* apparently referring to this story as it was handed down to him when he declares '*If Cain's revenge is sevenfold, then Lamech's is seventy-sevenfold*', presumably at his own hand.

Although no indication is provided as to how long the sons of Lamech lived, it is highly unlikely that Tubal-Cain could have been a metal-worker, since manufacture of goods from copper, as far as we can ascertain at present, did not commence until around 4500 BC and the Bronze Age commenced about a thousand years after that. The earliest indications of iron production are found in Turkey and date from around 2200-2000 BC. Therefore, if there is any genuine link between Tubal-Cain and metal-working, as the story reached Moses, it would only be on the basis that early metal-workers arose from the line of Tubal-Cain. I don't believe that much credence can be given to that claim, or (in verses 20 and 21) that harp and lute players all descended from Jubal or that all tent-dwellers were Jabal's descendants. However, these stories are of no consequence in the plan of God.

The *account of the heavens and the earth when they were created* ends in Gen. 4:25-26 with the line of descent from Adam through Seth. In Gen.

5:5, it is stated that Adam subsequently had other sons and daughters (Eve is not mentioned and at least some of these may have been from other wives). However, the line of Seth is the only line of descent that matters—it is the line through which God will continue to reveal himself to the world.

The most significant sentence in this chapter is the last: '*At that time men began to call on the name of the Lord*'. The 'time' referred to is the time of Seth and his son Enosh and suggests that God is no longer with Adam's descendants as a physical presence. The Hebrew word *qara*, here translated as 'to call', can also mean 'to proclaim'. The first alternative suggests a 'calling out to' in a situation where the caller believes the one called to can hear even if they cannot be seen. There is no indication here that there is a special place where the caller needs to go to be heard, although it is reasonable to think that the caller would probably go to the area where the physical presence was last encountered.

If the alternative translation, 'to proclaim', was intended, it suggests that Seth's family could have been sharing their experience of and faith in God with other settlers in the area. This would seem to be a reasonable thing on their part and would support the suggestion that 'men' began to accept the existence of an all-powerful god to whom they could turn for help.

10

ADAM TO NOAH

The remainder of this book is concerned with chapters 5-50 of *Genesis*. It is not intended to be a theological commentary. Instead, it tries to place the Genesis stories within the context of our scientific understanding of the world of those times. Therefore I will concentrate on theological issues only inasmuch as science can throw light on them or where they represent progressive revelations regarding the nature and purposes of God and humankind's relationship with him.

Our understanding of the times is heavily dependent upon the sciences of archaeology, climatology and geology. In the past century, archaeological research in the Levant, Mesopotamia and Egypt has enabled us to build a fairly cohesive knowledge of the Neolithic, Pottery and Bronze Ages in the area, but there are still many gaps and we can only make educated guesses about times and places. However, this research continues apace and each new discovery may help us to more accurately date events and throw more light on persons, races and cultures in the years to come. We need to keep this in mind as we investigate the *Genesis* accounts.

The Sources of Genesis

Apart from the *Elohim* account of the creation of the cosmos, the remainder of the *Book of Genesis* appears to have been compiled by the author from a series of ten pre-existing documents which together traced the history of the Children of Israel from their origin up to their settlement in the land of Egypt. The ten accounts are:

1. *'The account of the heavens and the earth when they were created'* (Gen 2:4-4:26)
2. *'The account of the generations of Adam'* (Gen 5:1-6:8)
3. *'The account of Noah'* (Gen 6:9-9:29)
4. *'The account of Shem, Ham and Japheth'* (Gen 10:1-11:9)
5. *'The account of Shem'* (Gen 11:10-26)
6. *'The account of Terah'* (Gen 11:27-25:11)
7. *'The account of Abraham's son Ishmael'* (Gen 25:12-18)
8. *'The account of Abraham's son Isaac'* (Gen 25:19-35:29)
9. *'The account of Esau'* (Gen 36:1-37:1)
10. *'The account of Jacob'* (Gen 37:2-50:26)

It is almost certain that at least the first two of these had existed wholly and the next three at least partly as memorised narratives for some time prior to their being written down (since they are believed to be of events that occurred prior to the development of writing).

Adam's Line

The account of Adam's line in Genesis 5 commences with Adam but is only concerned with his third son, Seth, and thereafter the first-born sons in the line from Seth to Noah and his three sons. This makes sense inasmuch as the later generations in the line no doubt thought that all other lines were wiped out by the great flood in the time of Noah and therefore could be disregarded. The account clearly existed in isolation from *'The account of the heavens and the earth'* since that account (in Gen 4:19-22) identifies Lamech's sons as being the 'fathers' of all those who lived in tents, played the flute and harp and forged tools out of bronze and iron.

The ages to which each of the named members of Adam's line up to and including Noah are given in Genesis 5, together with the age of each father when his first son was born. If we accept these ages as stated, this enables us to put together a 'time-line' from Adam's 'birth', which we will identify as Year '0'. If we then take the same details for Shem's line taken from *'The account of Shem'*, we can then extend this time-line to Abram (Abraham).

	Age at Death (yrs)	Lived From	Lived To
Adam	930	0	930
Seth	912	130	1042
Enosh	905	235	1140
Kenan	910	325	1235
Mahalalel	895	395	1290
Jared	962	460	1422
Enoch	365	622	987
Methuselah	969	687	1656
Lamech	777	874	1651
Noah	950	1056	2006
Shem	600	1558	2158
THE FLOOD		**1656**	
Arphaxad	438	1658	2061
Shelah	433	1693	2096

Eber	464	1723	2187
Peleg	239	1757	1996
Reu	239	1787	2026
Serug	230	1819	2049
Nahor	148	1849	1997
Terah	205	1878	2083
Abraham	175	1948	2123

If we were to take these figures as accurate, then less than 2,000 years would have passed between Adam's encounter with God and the birth of Abraham. Based on the later history of the Children of Israel when writing is widespread and we have a greater confidence in the timing of world events, we can reasonably date Abraham's life to the last two hundred years of the 3rd Millennium BC (2200-2000 BC). Counting back from this period would seem to date the Garden of Eden as late as around 4200 BC and the Flood around 2500 BC. This seems much too late in history given our knowledge of the civilisations of that time and conflicts with the Sumerian *King Lists*, which places the Flood some time before this period. It is clear that we have to look more critically at the figures provided in these accounts.

The Ages of Adam's Line

We know from current studies amongst long-lived peoples in the world that environment, diet, climate and physical activity all have an influence on how long we live; an inordinate number of Japanese seem to make the list of 'World's Oldest Living Persons'. However, despite current advances in diet and medical support, few humans seem to make it to the 120 year limit that God declared for mankind in Gen.6:3. Archaeologists have determined that stone-age peoples by and large were lucky to make it to half of that age (enthusiasts of the 'paleo' diet take note). So when we see people reported as having lived for hundreds of years, we need to be sceptical about three issues: how time was measured, how it was recorded and how it was tallied (counted).

Genesis is not alone in documenting long-lived characters back in those times—the *King Lists*, a list of Sumerian kings that has been constructed from various clay tablets and documents dating from the late 3rd Millennium BC (about the time of Abraham), attempts to provide a complete list of these kings dating back to times before the Great Flood. The eight antediluvian rulers (that is, rulers who lived prior to the flood) are reported to have each ruled for between 18,000 and 36,000 years, while the kings that ruled after the Flood did so for periods from 140 to 1500 years.

The first issue for considerations relates to the time periods that have been recorded as *years*. For thousands of years, primitive groups around the world had been aware of and tracked the cycle of the seasons relative to changes in the course of the sun's movements across the sky; as a result, they recognised a solar cycle of around 365 days. The Hebrew word *shaneh*, translated as 'year', is used consistently throughout the whole of the Old Testament to refer to a solar year and the compiler of *Genesis* would have accepted the ages in the original documents as meaning just that.

The second issue for consideration is how accurately and completely these ages were recorded. Of course, we should recognise the possibility that God could have ensured that these stories were accurately preserved and also that God did allow Adam's line to live extravagantly long lives. While this gives rise to difficulties in placing Adam's line within our current understanding of ancient history, it may find some support in its internal consistency—based on the previous chronological table, all of Noah's forefathers did not last beyond the Flood. In fact, Methuselah died in the year of the Flood; presumably either God waited until Methuselah died before he sent the flood or he perished in the Flood.

However, as I have already stated, it is almost certain that *'The account of the heavens and the earth'* and at least the early parts of *'The account of Shem'* began as verbal histories well before writing was invented around 3500 BC. Despite the best attempts of those charged with the responsibility for memorising and passing on details from generation to generation, it is improbable that this could have been performed without any errors, the most likely being the accidental (or intentional) dropping of people from the lists and the confusing of ages. We have no way of knowing whether such errors were introduced or how significant they might have been.

The third issue is probably the most significant—that of tallying (counting). Prior to the development of a written language there was no way to express numbers except by making a mark for each instance of items being counted; to convey numbers would have been limited to showing the number of fingers that matched the count and any number over that would effectively been classified as 'many'. Eventually, the ordinal numbers 1-9 would have been given names, together with a name for 10. By the time high-volume trade developed in the early 4th Millennium BC, a seller would incise on the outside of the container in which the goods were transported a separate mark for each of the items inside. Without an agreed way of summarising the quantity, the receiver then had to match each item of the goods with an incised mark to ensure that nothing had been lost or under-supplied. This practice was fairly quickly modified by grouping items into a kind of shorthand where a single symbol was used to represent a specific group size.

By around 3400 BC, the Sumerians had developed a number system where *sosses* were groups of 60 units, *ners* were 600 units and *sars* were 3,600 units. It is not clear why this numbering system was selected, but its influence is still felt today in respect to time and rotational units (60 seconds in one minute, 60 minutes in one hour, 360 rotational degrees in a circle and each rotational degree is made up of 60 minutes or 3600 seconds of arc). The lengths of reign by the antediluvian Mesopotamian kings were expressed in these units; the reign of *Ubara-Tutu*, last and shortest of these reigns is said to have lasted for five *sars* and one *ner*, equalling 18,600 units (years?). It was not until about 3100 BC that the Egyptians developed the 'system 10' numbering that we use today where the position of a numeral indicates whether it represents units, tens, hundreds, thousands, etc.). It was this system that Moses would have been educated in as part of '*all the wisdom of the Egyptians*' (Acts 7:22)

Many people may be offended by the suggestion that the *Genesis* accounts of Adam's line may be incomplete or inaccurate, but the above considerations must certainly give us pause for thought. As God's chosen people, the Jews have always placed great importance on the lineage of their race in their dealings with God and of their own individual lineage within it; their pride in that lineage is understandable. However, the significance of the history of the Children of Israel is not in the completeness of their genealogical records or the accuracy of their recorded ages—it is entirely dependent on the way that God has interacted with individuals within that race and what he has revealed of himself to all people through those interactions. Few even of those of Adam's line listed here have anything to teach us, so let us move on to more constructive and instructive matters.

11

SCIENCE AND THE FLOOD

In the whole of the *Book of Genesis*, no story (other than the creation of man himself) is as controversial with regard to its historicity as that of Noah and the Flood. In the scope of the God's revelation of himself to mankind, it appears to be a small matter and we could pass over it as a sideshow to the wider story. However, the importance of its symbolism and the doctrinal and scientific issues that it raises makes it worth devoting some time to.

Was There Ever a Worldwide Flood?

Stories of a great flood can be found amongst primitive peoples on almost every continent and most involve the survival of an individual or small group forced to start again. This fact is often used by some to support the idea of a worldwide flood which, in order to be large enough to affect our whole planet, must have wiped out most life on earth. It isn't surprising that such stories should be found; of all types of catastrophic events—volcanic eruptions, earthquakes, bushfires, famines, plagues and floods—floods are the most widespread and destructive calamities to occur in most regions on earth, whether it be in the Indus Valley, or along the Amazon or any of the large North American or European rivers or in the Middle East.

Unfortunately for the proponents of a single worldwide flood, no-one has ever been able to find any scientific evidence of such a flood; instances of large-scale flooding have been identified in various parts of the world, but they date to different times in the world's pre-history and are limited to areas of no greater than a few thousand square miles.

Some American fundamentalist Christians, very faithful, well-meaning Bible believers, claim that the Grand Canyon in itself is evidence of a mighty deluge. There we have thousands of feet of sedimentary rocks that forming the walls of the canyon. They maintain that only a flood of unbelievable dimensions could have deposited so much sediment and then, in draining away, could have cut such a large and spectacular canyon. However, they largely ignore what any photograph makes clear—the canyon displays not a single sedimentary layer but multiple layers.

Geological and chemical studies of the Grand Canyon have established that the Colorado Plateau in which the canyon is located consists of numerous sedimentary layers. The lowest of these that the Colorado River has exposed is the Vishnu Schist which has been dated at about two billion years old; the topmost layer is limestone that dates to about 230 million years old. Over billions of years, what is now the North American continent has been geologically active, various parts having risen and fallen. For much of the plateau's history it was actually lowland and, as sea levels rose and fell, it spent some periods as part of shallow seabeds, some as swamps and yet others as beaches or other near-shore environments. In each of these situations, sedimentary formations were laid down indicative of that type of environment. Between about one and a half billion years ago and half a billion years ago, the area was high and dry enough that negligible layer formation took place, so that there is an difference of a billion years between the 1.5 billion year-old layer and the 500 million year-old layer immediately above it.

Finally, starting about 65 million years ago, intense tectonic activity lifted the entire plateau 1,500-3,000 metres (5,000-10,000 feet). During this uplifting, the plateau endured twisting actions that created weaknesses in the plateau's structure, allowing the infant Colorado River and its tributaries to begin the task of carving out the canyon that is now one of the natural wonders of the world.

Was There a Great Flood?

In interpreting any of the tales of 'the flood' from anywhere in the world, we need to be conscious of the understanding of the cultures involved as to what constituted their 'world'. For example, the Hebrew word *erets*, translated in the account of Noah as 'the earth' or 'the world' can have the meanings: (physical) *earth, world, country, territory, district, region* and (figurative) *all occupants of*, depending on the context. Contextually, in Hebrew or any of the Mesopotamian languages of the period, the concept of the physical world was limited to the regions in which they lived or with which they were familiar through the trade of goods.

If we think within this more limited geographic scale then, four separate accounts of a Great Flood in the region of Mesopotamia have been found amongst ancient (mostly fragmented) records. These are (with the approximate dates of creation of the main fragments):

- The Ziusudra account (17th Century BC)
- The Epic of Gilgamesh (2150-1400 BC)
- The story of Atrahasis (early 2nd Millennium BC)
- The account of Noah (included in *Genesis*, probably mid-2nd Millennium BC).

All four sets of records refer to events from a much earlier time and are presumably copied from or based on earlier versions of the tales. All stories excepting the Epic of Gilgamesh are remarkably similar inasmuch as each tells of God (in the case of Noah) or the gods of having become exasperated by humankind's bad behaviour and deciding to destroy them. But God (in the case of Noah) or one of the gods warns a single wise or righteous individual (Noah, Atrahasis or Ziusudra) about the intended destruction and instructs the man to build a ship to save himself, his family and two of each kind of animal. In the Atrahasis account, Atrahasis floats down into the Persian Gulf; Noah and Ziusudra finally come to rest on mountains. In the Gilgamesh story, Gilgamesh seeks immortality by searching for Utnapishtim, an ancient man who had survived a great flood because, like Ziusudra, Atrahasis and Noah, he was warned by a god to build a boat for the saving of his family and animals of every type.

The similarity of the tales strongly suggests that they have a common origin, possibly referring to the same event or even the same individual. We have no idea as to which might be the most accurate or original version; clearly, each version is structured to suit the culture and religious beliefs of the writer, regardless of the source of the original tale. The account of Noah is concerned with Adam's line to Abraham and so is included in *Genesis* as being directly pertinent to the history of the Children of Israel.

A careful reading of Genesis 6:1—9:17 suggests that the story of Noah and the flood may have been compiled from more than one source:

- Within the text, the Hebrew *Yehovah* (the Lord) is used ten times and *Elohim* (God) is used thirteen times. These seem to occur almost randomly but in fact tend to 'clump', as if the author of *Genesis* has been doing a 'cut and paste' job;

- In Genesis 6:20-21 and again in 7:14-16 Noah is instructed to take two of every kind of clean and unclean animal, bird and 'creepy crawly' (a male and a female in each case) into the ark. However, in 7:2-3, God specifies *'seven pairs of all clean animals, the male and his mate, and a pair of the animals that are not clean, the male and his mate, and seven pairs of the birds of the heavens also, male and female'*.

Locating and Dating the Flood

As might be expected, there appear to have been numerous floods within the territories through which the Tigris and Euphrates Rivers pass. However, one particular flood event around the city of Shuruppak (modern Tell Fara in Iraq) seems to have been most extensive, reaching as far as Kish. Kish is a city kingdom which is recorded (in the King Lists and other records) as having taken over leadership of the whole of the region after that flood, presumably because the other city kingdoms more exposed to the flooding had been devastated. It is thought that the Shuruppak flood was the result of spreading dunes damming the Karun River, which flooded into the Tigris while, at the same time, heavy rainfall in the Nineveh region added to the volume of water and the resulting torrent spilled across into the Euphrates.

Pottery from the Jemdet Nasr period (c. 3000–2900 BC) was discovered immediately below the Shuruppak flood stratum and the sediments from the flood itself have been carbon-dated to about 2900 BC.

Archbishop Usher, using the ages given for the various members of Adam's line given in *Genesis* together with the best known historical dates available in his time, dated the Flood at 2349 BC. If we accept that Abraham actually lived from about 2200 BC to about 2000 BC and using the timeline provided back in Chapter 10, the Flood would be more accurately dated to about 2600 BC. Both of these dates, although based on unreliable data, would seem to suggest that 2900 BC could be a very reasonable date for 'Noah's flood'.

Two other possibilities have been suggested as possible bases for the flood stories:

- A meteor or comet crashed into the Indian Ocean around 3000-2800 BC, creating the 30-kilometre (19-mile) wide undersea Burkle Crater. It is suggested that the resulting tidal wave could have devastated coastal lands. To date, no indication of such flooding has been found in the Persian Gulf and this suggestion can probably be discounted.
- Around 5600 BC, the sea level of the Mediterranean reached a point at which the pressure of the water breached a rocky sill in the Bosporus Strait, causing a catastrophic rise in the level of the Black Sea. It is suggested that Noah might have been living in that region at the time, rather than in Mesopotamia, and that this was the flood that he survived. This alternative event has the advantages of being much closer to Mount Ararat, where the ark is said to have finally grounded and fits more realistically with the time that the ark is supposed to have been afloat before coming to rest. While scientists agree that this event occurred, it is by no means clear as to the speed with which it occurred or the extent of the impact on the inhabitants. If it was the basis of the Noah version of the Flood, it would separate it from the other three Mesopotamia-based tales.

Noah's Ark

The oldest man-made craft believed to have been used to travel over water were rafts constructed by binding multiple pieces of timber together. We have no idea how long ago such craft were developed, but it could even be that they precede the emergence of Homo sapiens. Dugout canoes required more advanced tools but were in use by the early 8th Millennium BC, if not considerably earlier. We also know that sea-going craft made of bundled reeds coated with pitch (naturally-occurring tar) were in use by around 5000 BC because the remains of one that age has been found on Failaka Island, Kuwait, located about 20 km off Kuwait City. However, a petroglyph found in Azerbaijan and which seems to depict a reed boat has been tentatively dated to around 10,000 BC.

Boats built from wooden planks, caulked with reeds and pitch, seem first to have been built in Egypt as early as 3100 BC. By 2600 BC, Egyptians had mastered the building and sailing of such ships and it is probable that this technology had spread to other countries skirting the Mediterranean Sea. The first named ship that historians know about is the *Praise of the Two Lands*, a large Egyptian ship built under the pharaoh Sneferu around 2600 B.C.

The so-called *Khufu ship* is believed to have belonged to Pharaoh Khufu of the old Egyptian kingdom. This perfectly preserved vessel was sealed in the Pyramid complex in Giza around 2500 BC and is about 143 feet long and 19.5 feet wide. It was never used as a ship, being intended for the pharaoh's use in the after-life.

Genesis 6:15 tells us that the ark was to be a wooden-planked boat, 300 cubits long, 50 cubits wide and 30 cubits high. The ancient cubit was based on the distance from a man's elbow to the tips of his fingers. Clearly, this was likely to result in variations from time to time and place to place. To date, all standard measurements of the cubit have been found to fall within the range of 17.5-20.8 inches.

In the 1985 version of the NIV Study Bible, the translators chose to use the later Hebrew value of the cubit as used in the building of the first temple which was equivalent to 18 inches. This gave them dimensions of the ark as 450 feet long by 75 feet wide by 45 feet high. If the story of Noah was indeed included in *Genesis* at the time of Moses, it is unlikely that these figures are correct. In Mesopotamia during the early 3rd Millennium BC, every kingdom and every trade guild had its own standard for the measure.

In 1916, in the middle of World War I, a copper-alloy bar was found during an excavation at Nippur in Iraq. The bar dates from around 2650 BC and seems to be a measurement standard; if so, it defined the Sumerian cubit of the time as about 518.6 mm (20.4 in). If this was close to the standard unit used at the time of building, then the ark's dimensions would have been 510 by 85 by 51 feet. If, on the other hand, the documents from which Moses copied the story had used the Royal Egyptian value for the cubit of 20.7 inches, the ark would have measured 517.5 by 86.25 by 51.75 feet. It is clear that the actual difference between the two sets of dimensions is insignificant. What *is* significant is that the length of the ark is over three and a half times the length of the *Khufu ship*, built perhaps some hundreds of years later and considerably longer than any other timber ship ever built. In modern terms, it was as long as one and a half soccer football fields!

On the basis of what we have already concluded about the Flood, timber-planked ships caulked with pitch were probably known about, if not used, in Mesopotamia at the time. We aren't told Noah's occupation, so it could just be that he was a ship-builder; otherwise he would have had to commission the building of the ark by those with professional experience in the industry. Aside from the planning, the actual building of the ark (including obtaining and preparing the required timber) would have taken many thousands of man-hours. The account is not clear how much warning Noah was given, but the task was clearly beyond the capability of Noah and his three sons working alone.

Manpower aside, there remains the problem of stability. Other than the ark, the largest sea-going wooden sailing ships have been the *Wyoming* (launched in 1909 and 450 feet long), and the *Columbus* (launched in 1824 and 356 feet long); both were built by master ship-builders but each notoriously suffered from twisting and buckling of the timber planks in heavy seas. The *Columbus* successfully sailed from the United States to England, but broke up and sank in the English Channel on its return voyage. The *Wyoming* had its timbers braced with steel and managed to sail the Great Lakes for around twenty years before it broke up and sank with the loss of all hands. It is unlikely that the engineering skills of boat-builders of Noah's time could have succeeded in building a vessel of its reputed size that could handle the maelstrom of the Flood, survive 150 days afloat and then avoid breaking its back as it grounded on Ararat while the waters swirled and subsided around it.

Despite all of the engineering problems faced by the builders of a boat this size, we cannot entirely discount God's miraculous protection of the ark throughout the flood nor God's ability to bring the required types and numbers of animals to the ark. However, the scientific conclusion that the flood, while probably extensive, was of regional rather than world significance suggests that the required number of animals could have been accommodated in a much smaller craft. Therefore, whatever historical basis there might have been for the Noah account, it seems that details of the ark's dimensions and how long the floodwaters covered the land were probably greatly exaggerated as they were passed down through the generations.

In recent times there have been many attempts to locate the remains of Noah's ark and these have resulted in a number of claims of success. Many of these can be found on YouTube. Unfortunately none have been able to withstand scientific investigation.

For readers who want to investigate the matter further, I would suggest watching an excellent YouTube video that can be found at:

https://youtu.be/8JRBKYdnhzY

12

BIBLICAL SIGNIFICANCE OF THE FLOOD

Living around the middle of the 2nd Millennium BC, Hebrews in the time of Moses almost certainly regarded the story of Noah and the Flood as history. As such, it played a critical part in establishing the lineage of the Children of Israel back to Adam. However, I believe that God intended its inclusion in *Genesis* for a much deeper purpose. Whatever scientific conjectures might be made about its historicity or the accuracy of many of the details within the text, our main concern in this book is to consider the part the story plays in revealing more about the nature of God and his plan for humankind.

While much of the significance of the account of Noah lies in its typography, there is also much that impinges directly on some the deepest theological controversies amongst the various Christian denominations. I will deal firstly with the references to Noah in the rest of the Bible and then with some theological issues in the order in which they arise in the story. I will deal with a specific major issue in the next chapter.

Later Biblical References to Noah and the Ark

Exodus 3:3, 5

The Hebrew word translated as 'ark' in *Genesis*, chapters 6-8 is *tebah*. The word refers to a container. The only other places this word appears in the Bible are Exodus 3:3-5, where it is translated as 'basket', in this case the container is woven from reeds. Like the ark in the story of Noah, it is made water-proof by lining it inside and out with pitch and tar.

The use of *tebah* in each case highlights that the vessel is used to preserve the life of God's chosen one, Noah in the first instance and Moses in the second.

Pitch is a bituminous material that oozes out of the ground in geothermal areas and hardens into tar. Both materials were widely used to water-proof containers and ships and can still be found occurring naturally near Babylon and around the southern end of the Dead Sea.

Isaiah 54:9

In this passage, God specifically refers to his covenant with Noah that 'the waters … would never again cover the earth (*country, territory, district, region*)', which would seem to affirm that there was a historical basis to the flood story.

Ezekiel 14:14-20

In these verses, God declares that his judgement on any country that sins and is unfaithful to him will be executed and that 'even if …Noah, Daniel and Job were in it, they could only save themselves by their righteousness'.

The emphasis here is on the universally recognised righteousness of these three individuals and not necessarily on their historicity.

Matthew 24:37-38, Luke 17:26-27

In both accounts of the same incident, Jesus likens the state of the world at the time of his return to that of the world at the time of the flood, with people living their lives without regard to the warnings of judgement that they had received.

Again, this statement refers only to the godlessness of the world of the time, which had become a by-word, and doesn't depend on its historical basis.

Hebrews 11:7

In the author's list of persons of faith, Noah is stated to have believed God and obeyed his command and, as a result, saved his family.

2 Peter 2:5

Peter compares the judgement awaiting those who deliberately lead believers astray to the judgement visited on the wickedness of the people of Noah's day. This reflects Jesus' own references to what had become a by-word to the Jews.

The Sons of God and the Daughters of Men

Genesis 6:1-4 reads:

> *¹ When man began to multiply on the face of the land and daughters were born to them,*
>
> *² the sons of God saw that the daughters of man were attractive. And they took as their wives any they chose.*

> *³ Then the LORD said, "My Spirit shall not abide in man forever, for he is flesh: his days shall be 120 years."*
>
> *⁴ The Nephilim were on the earth in those days, and also afterward, when the sons of God came in to the daughters of man and they bore children to them. These were the mighty men who were of old, the men of renown.*

This passage introduces three groups who have become a centre of controversy since early Christian times—the daughters of men, the sons of God and the Nephilim. The key group is the one termed *the sons of God*. The controversy arises because there are two possible interpretations of the Hebrew phrase *ben Elohim*, translated literally as *sons of God*.

- The Hebrew phrase occurs in the Old Testament only in this passage and in the Book of Job. In Job, the phrase clearly refers to angels. For this reason, the same phrase in Genesis 6 was commonly inferred to also refer to angels.
- The phrase may also specifically refer to the Sethites—that is, the descendants of Seth that accepted and passed on their knowledge of the One True God, '*YHWH*' or *Yehovah*.

Most of the earliest Jewish sources from the 3rd Century BC accepted the idea that Genesis 6 referred to angels and some of the earliest Christian leaders such as Justin Martyr, Eusebius, Clement of Alexandria and Origen subscribed to this view. However, by the 2nd Century AD, the preponderance of Jewish rabbinic teachers and Christian leaders such as Augustine of Hippo, John Chrysostom, and, much later, John Calvin have accepted the belief that the group mentioned in Genesis 6 actually referred to the descendants of Seth; only groups such as the Jewish sect, *the Essenes*, centred around the Dead Sea area, clung to the angelic interpretation as found in the *Dead Sea Scrolls* and the apocryphal books of *1 Enoch*, *2 Baruch*, *Jubilees* and the *Testament of Reuben* (For example, 1 Enoch 7:2 says '*And when the angels, the sons of heaven, beheld them, they became enamored of them, saying to each other, Come, let us select for ourselves wives from the progeny of men, and let us beget children*').

A major influence on the early Christian writers who rejected the angel hypothesis was Christ's statement in Matthew 22:30 that angels do not marry (although some writers held that this referred only to angels in heaven, not the fallen angels).

From a scientific point of view, while we might accept that angels as messengers of God might be able to take on human form, it is unlikely that they would also have the genetic make-up that would incline them to want to marry and enable them to procreate with human women. On this basis then, it would seem far more likely that the descendants of Seth would think of themselves as 'the children of God' and even that the neighbouring peoples would refer to them as such.

If this so, then 'the daughters of men' most likely refers to peoples outside the *Sethites* and not to humankind in general. It is likely that intermarriage with other peoples in the region led to the *Sethites* becoming willing participants in the great wickedness described in Genesis 6:5. (We should remember that, in Moses' time, God specifically commanded the Children of Israel not to intermarry with the peoples around them when they moved into the Promised Land because they would then be led astray to worship other gods.)

In passing, it is worth noting that if God was intent on only destroying Adam's line because they were the ones that had turned their backs upon God, a large-scale Mesopotamian flood would probably have been sufficient and is much more in keeping with the scientific facts that there was no world-wide flood and that humans by that time had spread to every continent except Antarctica.

The greatest disagreements between scholars over these few verses in Genesis 6 are in relation to the third group. The identity of the *Nephilim* is complicated by the fact that the Hebrew language of the time makes it unclear whether it is *the sons of God* themselves who were the *Nephilim* or it was their offspring or whether, in fact, they were a separate group altogether. The lack of vowels in written Hebrew has led to some believing that the word should be nophlim, which translates to 'fallen' and that the reference is to the fallen angels.

In the King James Version of the Bible and in many other English translations since then the Hebrew word *nephiyl* is translated as 'giants', while the remaining English translations simply call them 'the Nephilim'. The idea that the Nephilim were giants arises from Numbers 13:32-33 where the spies sent out by Moses to assess the land of Canaan report that '*all the people that we saw in it are of great height. And there we saw the Nephilim (the sons of Anak, who come from the Nephilim), and we seemed to ourselves like grasshoppers, and so we seemed to them*'. While it is possible that the Nephilim were indeed a genetically-related group of giants, there is no real likelihood that they arose from sexual relations between angels and human women.

If we consider the Hebrew word *nephiyl* itself, its root construction implies tyranny, bullying, or attacking (falling upon). On that basis, my own interpretation is that the Nephilim were a third, contemporaneous group of *gibbowr* (mighty, brave, strong, warriors) that used aggressive and violent behaviour to further their own ends (or the ends of those who hired them) and significantly contributed to the wickedness and violence that brought God's judgement on mankind. The fact that many of these became *'iyst shem* (renowned or famed individuals) seems to back this up. However, I am not in any way an expert in the Hebrew language—Biblical or modern.

The Limit on Human Lifespans

Earlier in this book I suggested that the lengthy lifespans quoted for Adam's line may have been exaggerated, particularly in the days before the development of man's ability to count to high numbers or to express those numbers in writing. Archaeological measurements of the ages of skeletons for the time and region certainly indicate that general lifespans were well under one hundred years. However, it is possible that Adam's line did enjoy an extended lifespan by the grace of God. Certainly, by Noah's time, numeration was sufficiently well advanced and ages could be more reliably recorded, yet Adam's line continued to display great longevity.

In Genesis 6:3, God decrees that the lifespan of men will be limited to 120 years. In the context of the surrounding verses, this makes sense if the *sons of God* were the descendants of Seth and relates to their intermarriage with the surrounding tribes. Nevertheless, *Genesis* reports that the principal line from Noah to Jacob lived beyond this limit.

In modern times, the only person who has been verified as having lived beyond 120 years was Jeanne Calment (21 February 1875—4 August 1997, aged 122 years and 164 days).

The Rainbow

In Genesis 9:8-16, God makes a covenant with Noah that never again would he destroy the 'earth' (region?) by a devastating flood. God then establishes the rainbow as a perpetual symbol of that covenant. There is no suggestion that rainbows did not exist prior to that time; rainbows have existed and will continue to exist whenever sunlight is refracted through droplets of water in the atmosphere. God is simply using the rainbow as a reminder to both himself and mankind that as long as rainbows exist, so too will that covenant.

13

PREDESTINATION AND FREE WILL

This subject might seem to be out of place in a book about a scientific view of Genesis, but in fact we are first confronted by this issue in Genesis 6:

> *⁶ And the LORD regretted that he had made man on the earth, and it grieved him to his heart. ⁷ So the LORD said, "I will blot out man whom I have created from the face of the land, man and animals and creeping things and birds of the heavens, for I am sorry that I have made them."*

Can God Make Mistakes?

In the Judeo-Christian culture in which most of us have been raised, we have been taught that God is omniscient (knowing everything) and omnipotent (able to do everything). We are appalled at the thought that God might not be in total control, might fail or might make mistakes, yet here we are told that God 'regretted' having done something.

And this is not the only passage in the Bible that implies this. In 1 Samuel 15:11, God tells Samuel: *"I regret that I have made Saul king, for he has turned back from following me and has not performed my commandments"*.

It was God who chose Saul to be king, so whose fault was it if Saul didn't fulfil God's expectations? God himself supplies the answer in speaking to Saul in verse 23 of the same chapter: *"For rebellion is as the sin of divination, and presumption is as iniquity and idolatry. Because you have rejected the word of the LORD, he has also rejected you from being king"*.

So it was Saul's disobedience that resulted in God's rejection of him as king. But why then, if God is all knowing, did God choose Saul in the first place? Isn't God ultimately at fault?

We could try to avoid the issue in the Genesis 6 verses by suggesting that the tale's author is merely interpreting God's feelings as if God himself was human. However, in 1 Samuel it is God speaking directly, first to Samuel and then to Saul.

If we are going to be completely honest in our appraisal of *Genesis*, we have to look at the implications of these Genesis 6 verses and those in 1 Samuel 15. In doing so, I believe that science can give us a new and deeper understanding of one of the most difficult and divisive doctrines that afflicts the Christian church today—the extent to which God exercises control over everything that happens in the universe.

But first we should look at how Bible scholars have looked at this question. This is not the time or place to go into the subject in great depth, but the following is a brief summary of the major positions.

Theological Approach

Theologically, there are three points of view:

- God's control is absolute—nothing has ever happened or will ever happen except under God's control and in accordance with his own will and plan. This view was promoted by John Calvin in the 16th Century AD and became known as *Calvinism*.

- God limits his control to allow man freedom to respond to God's will. This view was put forward by Jacobus Arminius some years later and is called *Arminianism*.
- Most Christian theologians accept that both appear to be taught in the Bible and so both must be true to some extent, even though they appear to be contradictory.

Calvinism

Strict Calvinistic doctrine insists that nothing can occur outside of God's predetermined plan; to allow anything outside of his specific plan would risk that plan being damaged or even failing completely. To ensure that God remains in control, Calvinists believe that there is no place for mankind to exercise free will; free will becomes an illusion—sinful mankind is incapable of responding to God through their own volition. Calvinists believe that before the creation of the world God chose (or 'elected') some to be saved and that election has nothing to do with man's future response; rather, the elect are really powerless to refuse the salvation that God offers just as the remainder of mankind is powerless to seek and receive the salvation that Jesus Christ brought about through his death and resurrection.

Many go so far as to believe that Christ died to save only those elected by God the Father before the world began. Therefore, since Christ did not die for everyone, but only for the elect, his atonement was wholly successful.

Central to Calvinism is the doctrine of 'Predestination'. This term comes from the Greek word *proorizo*, meaning 'to decide beforehand' or 'appoint beforehand'. The word appears only six times in the New Testament—not always translated the same way:

> *For truly in this city there were gathered together against your holy servant Jesus, whom you anointed, both Herod and*

> *Pontius Pilate, along with the Gentiles and the peoples of Israel, to do whatever your hand and your plan had **predestined** to take place. (Acts 4:28)*
>
> *For those whom he foreknew he also **predestined** to be conformed to the image of his Son, in order that he might be the firstborn among many brothers. And those whom he **predestined** he also called, and those whom he called he also justified, and those whom he justified he also glorified. (Romans 8:29-30)*
>
> *But we impart a secret and hidden wisdom of God, which God **decreed before** the ages for our glory. (1 Corinthians 2:7)*
>
> *He **predestined** us for adoption to himself as sons through Jesus Christ, according to the purpose of his will. (Ephesians 1:5)*
>
> *In him we have obtained an inheritance, having been **predestined** according to the purpose of him who works all things according to the counsel of his will. (Ephesians 1:11)*

Calvinists infer from these verses that God has to exercise complete control over his world to ensure that his predetermined plan for mankind is achieved and that every prophecy is successfully and accurately fulfilled.

There are a number of unpleasant corollaries to this doctrine, some of which are:

- All mankind are just pawns in a game that God is playing.
- Every human is born as an unredeemable sinner and is destined for hell or heaven solely at God's whim.

- While man's sinfulness may be the cause of all the evil and suffering in the world, God must bear ultimate responsibility for the situation since he is the only one with the power to correct it. The belief that God is wholly good, wholly merciful and wholly loving is thus turned into a lie.
- While God himself states that he promotes, incites and uses some evil men in the fulfilment of his purposes, strict Calvinism requires that every monster of the likes of Adolf Hitler, Josef Stalin and Mao Tse-tung gained power specifically in accordance with God's will and plan and therefore that God is ultimately responsible for the horrors that they committed.

While Calvinists might reject these accusations, they appear to be logical outcomes derived from the Calvinist stance.

Arminianism

The Arminian approach to scripture recognises God's ability and right to control but believes that he specifically limits that power to allow mankind to exercise free will in their response to him and in their continued relationship with him. Man's right and responsibility to choose is stated throughout the whole Bible, as for example:

> *"I call heaven and earth to witness against you today, that I have set before you life and death, blessing and curse. Therefore choose life, that you and your offspring may live". (Deuteronomy 30:19)*

> *"And if it is evil in your eyes to serve the LORD, choose this day whom you will serve, whether the gods your fathers served in the region beyond the River, or the gods of the Amorites in*

> *whose land you dwell. But as for me and my house, we will serve the LORD". (Joshua 24:15)*

While verses such as these relate directly to making choices, there are literally hundreds of verses that tell us how we are required to live; such requirements only makes sense if we have the free will to obey or disobey.

The Arminian point of view also recognises that God has a plan that has been in place from the beginning and that this plan of redemption involved Jesus dying for our sins. While it was all part of God's plan for Jesus to die, his arrest, trial and crucifixion were the direct result of deliberate human actions:

> *He was foreknown before the foundation of the world but was made manifest in the last times for the sake of you. (1Peter 1:20)*

> *"Now is my soul troubled. And what shall I say? 'Father, save me from this hour'? But for this purpose I have come to this hour'. (John 12:27)*

> *"This Jesus, delivered up according to the definite plan and foreknowledge of God, you crucified and killed by the hands of lawless men". (Acts 2:23)*

Arminians also recognise that God chooses people as agents for the bringing about the fulfilment of his plan. As examples, God chose Noah (Genesis 7:1), Abraham (Genesis 18:19), Bezalel (Exodus 35:30), the Children of Israel (Deuteronomy 7:6), Saul (1 Samuel 10:24), David (1 Samuel 16:1, 12) and Saul of Tarsus (Acts 9:15). God even uses bad people to bring about his own purposes and glory; for example, he says to Pharaoh:

> "But for this purpose I have raised you up, to show you my power, so that my name may be proclaimed in all the earth". (Exodus 9:16)

The implication of this verse is not that God *made* Pharaoh behave as he did, but rather God ordered events so that Pharaoh, being the kind of person he was, was in a position to act as he did.

With regard to whether Jesus' death was limited in its effectiveness, the New Testament writers make it abundantly clear that Christ died for all humankind and that God's desire is for all to come to him in faith, as the following verses attest:

> *For God so loved the world, that he gave his only Son, that **whoever** believes in him should not perish but have eternal life.* (John 3:16)

> *For the death he died he died to sin, once **for all**, but the life he lives he lives to God.* (Romans 6:10)

> *And he died **for all**, that those who live might no longer live for themselves but for him who for their sake died and was raised.* (2 Corinthians 5:15)

> *The Lord is not slow to fulfill his promise as some count slowness, but is patient toward you, **not wishing that any should perish**, but that **all** should reach repentance.* (2 Peter 3:9)

> *He has no need, like those high priests, to offer sacrifices daily, first for his own sins and then for those of the people, since he did this once **for all** when he offered up himself.* (Hebrews 7:27)

> *But when Christ appeared as a high priest of the good things that have come, then through the greater and more perfect tent (not made with hands, that is, not of this creation) he entered once **for all** into the holy places, not by means of the blood of goats and calves but by means of his own blood, thus securing an eternal redemption. (Hebrews 9:11-12)*

> *And by that will we have been sanctified through the offering of the body of Jesus Christ once **for all**. (Hebrews 10:10)*

Can We Have It Both Ways?

Bible verses can be found to support both theological positions, so the debate continues. Theologians disagree about the weight to be given to the various verses or which is the correct interpretation to be placed on some verses. Aside from some extreme Calvinist and extreme Arminian groups, most denominations hold to a middle-of-the-road acceptance of the truth of both points of view and believe that we will never know in this life something that is beyond our human ability to understand fully.

To a scientist, this is an unsatisfactory state of affairs; science needs to find answers. It is all a part of 'thinking God's thoughts after him'. So, how can we begin to resolve the issue? I believe the answer must start with understanding what constitutes *the foreknowledge of God*.

There are two passages in the New Testament that are pivotal to an understanding of predestination:

> *For those whom **he foreknew** he also predestined to be conformed to the image of his Son, in order that he might be the firstborn among many brothers. And those whom he predestined he also called, and those whom he called he*

> also justified, and those whom he justified he also glorified. (Romans 8:29-30)
>
> Peter, an apostle of Jesus Christ, To those who are elect exiles of the Dispersion in Pontus, Galatia, Cappadocia, Asia, and Bithynia, **according to the foreknowledge of God** the Father, in the sanctification of the Spirit, for obedience to Jesus Christ and for sprinkling with his blood: May grace and peace be multiplied to you. (1Peter 1:1-2)

Depending on the inclination of the theologian, the *foreknowledge of God* can be taken to mean either that God decided in advance who was to do what in accordance with his specific plan and that each person will inevitably fulfil his/her role (the Calvinist belief) or that, because God knows in advance who are going to accept him, he calls them to be part of his kingdom and to participate in the fulfilment of his plan (the Arminian approach). This is the point where science can shed new light.

The Scientific View

God, Space and Time

God is separate from the universe he created, so he is not bound by space and time as we know them because they are properties of this universe. The universe came into existence and will eventually grow cold and lifeless; God is eternal. We see the universe in terms of near and far, large and small, past, present and future; for God, the entire history of the universe is *present reality*. For God, every point in the universe is 'here' and every moment in time is 'now'. This why Jesus could say *"Are not two sparrows sold for a penny?*

And not one of them will fall to the ground apart from your Father" (Matthew 10:29) and *"Truly, truly, I say to you, before Abraham was, I am"* (John 8:58).

We must not think of God's view of time in the universe as if it was linear; it is simultaneous. This means that every past, present or future natural occurrence in the universe or on Earth, every action by man or beast and every intervention by God is happening now in God's present and in God's presence. The question then is to what extent does this place everything under God's direct control?

Degrees of Freedom

God has built into this universe many degrees of freedom. The creation of the universe was cataclysmic. It has evolved through chaotic processes whereby stars form from the debris of former stars, then they blaze and die—either by quietly burning out or by exploding as giant supernovae. Galaxies collide; black holes swallow whole stars and their associated planets. The universe has untold numbers of objects of various sizes doomed to travel through space until they collide with, or are captured by, larger heavenly bodies. Earth is still occasionally subject to astronomical catastrophes such as the asteroid or comet that crashed into the Yucatan Peninsula in south-east Mexico almost 66 million years ago, wiping out 75 per cent of all plant and animal life on the planet. We should not think of these things as individually planned acts of God—they are random phenomena that are vital parts of the very nature of the universe. Earth is part of that natural evolution. The very composition and structure of our planet means that we are subject to random phenomena such as earthquakes, tsunamis, volcanic eruptions, periods of disastrous flooding and droughts.

Life itself exhibits its own degrees of freedom. Each species has its own life cycle and occupies its own niche within its general environment. But individual creatures are subject to forces and events outside of their control—predation, conflicts, accidents, diseases and natural disasters—while attempting to follow their own instincts to survive and maintain the species. Nothing is certain except inevitable death. Homo sapiens is one of those species. We would be very foolish to think that God is so intent on ensuring the fulfilment of his plan that he would go to the extreme of controlling the history of every individual creature on the planet, past, present and future.

Chaos Theory and the Butterfly Effect

Chaos theory is an area of mathematics that tries to understand and model the behaviour of complex systems in an effort to accurately predict the outcomes when starting from known initial conditions. The most easily recognised example relates the effects of air temperature, pressure and humidity, together with topography of a geographical region to forecast the weather. The frequent failures of such models to accurately predict the weather arise from three things: the difficulty involved in identifying all of the variables involved, the difficulty in correctly quantifying the relationships between these variables and the difficulty in accurately determining the initial values of each of the variables involved. Advances in computing power, together with better identification of the variables and their quantitative effects, have generally improved weather predictability, but still leave much to be desired.

When a system had been refined to the extent that all variables and their interrelationships are fully understood, these systems are said to be deterministic, meaning that their future behaviour is fully determined by their initial conditions, with no random elements involved. Even then, the deterministic nature of these systems does not make them predictable. This behaviour is known as *deterministic chaos* or simply *chaos* and occurs because the initial conditions can never be known to the necessary level of accuracy. One of the leading lights of Chaos Theory, Edward Lorenz, expressed the theory as:

Chaos: When the present determines the future, but the approximate present does not approximately determine the future.

The accurate determination of initial conditions can be virtually impossible to achieve because of the so-called *Butterfly Effect*. This effect describes how a small difference in one variable of a complex system can result in large differences in a later state of that system. The name given to this behaviour is based on the idea that the air disturbance caused by a butterfly flapping its wings in a Brazilian forest can become a contributory cause of a hurricane in Texas. This seems extreme but it is true, even if the effect is microscopically small relative to all of the other contributory effects.

On the basis of the above, it is clear that giving mankind free will in no way complicates God's plan any more than do the other degrees of freedom that are already operating in nature.

The Foreknowledge of God

Returning to the idea that God is not bound by time but is eternally *present*, we can now begin to look at the real meaning of *the foreknowledge of God*. It may be the hardest thing that we could ever be asked to understand with our finite minds.

It must start with the realisation that God sees the entire history of the universe, not as sequence of events but as a simultaneous whole. The first and most important implication of this is that God sees only what has been and will be—there are no 'might have beens'; things will take their natural course unless God intervenes at specific points (supposing that he chooses to). If, at some point in (our) time he intervenes to change some circumstances, that intervention becomes part of history; what the ultimate results might have been if he had not intervened remain unknown to God and to man. If he fails to intervene, that failure to intervene becomes part of history; he (and we) will never know what might have happened if he had intervened.

There a series of supplementary implications arising from this scientific fact. The first is that every prophecy that appears in the Bible must be true because its fulfilment is known to God as part of his *present*; he doesn't have to manipulate future history to make it come true.

The second implication is that, while God might know our innermost thoughts and desires better than we might know them ourselves, he has granted us free wills and thus has placed himself in a position where he can be disappointed or angry when we fail to live up to his hopes and expectations. This is why God 'regretted' that he had ever made man in view of the resultant mass failure of man to live up to even the basic impulses of the moral law in Noah's time. It is why he regretted making Saul king.

Fulfilling God's Plan

It is worth pursuing the story of Saul further, as it appears in 1 Samuel, chapters 8 to 13.

Israel wanted a king. God chose Saul on the basis of the kind of person Saul was—humble, tall, courageous, the kind of person that people would happily follow; he appeared well suited to the role.

He started out well, saving the men of Jabesh Gilead from the Ammonites, but then became impatient, disobeying God's commands and acting according to his own lights. It is instructive to look at what Samuel has to say to Saul in 1 Samuel 13:13-14:

> *And Samuel said to Saul, "You have done foolishly. You have not kept the command of the LORD your God, with which he commanded you. For then the LORD would have established*

> *your kingdom over Israel forever. But now your kingdom shall not continue. The LORD has sought out a man after his own heart, and the LORD has commanded him to be prince over his people, because you have not kept what the LORD commanded you."*

Samuel clearly states that if Saul had continued to obey God, God would have established Saul's kingdom over Israel for all time. That was God's intention. Did God's plan fail?

No. If Saul had continued to obey God, it would be Saul's line from the tribe of Benjamin through which Christ would have come. We might never have heard anything about David. Every prophecy would have been centred on Saul's descendants, because that would have become the historical reality. You see, the heart of God's plan was that he would send his son to be the Saviour of the world. His ultimate intention was based on the Messiah, not on the Messiah's ancestry. If God's plan is hindered at any point, God will find a way around the hindrance and the plan will go on.

Any military leader will tell you that ultimate victory depends on three things:

- A goal—what is to be achieved and how we will know when we get there;
- Strategies—the general means by which the goal is to be achieved based on our strengths and the enemies perceived weaknesses. They largely determine where our resources are to be concentrated and applied;
- Tactics—the day-by-day planning and execution of actions to achieve those strategies. Tactics are unpredictable because they largely lie outside of our control (a case in point being of an army

officer assigning a company of men the task of capturing a hill; whether they succeed will depend on how strongly and cleverly the men fight and on how well the hill is defended).

God's plan can never be foiled by opposition from any evil powers or failure on our part (although our failures might result in some individuals for whom Christ died never having the chance to hear and believe the gospel message). It is exciting to think that we can play a part in the fulfilment of God's plan and reassuring that our inadequacies can never cause it to fail. God will always have alternatives.

I think it was Archbishop Desmond Tutu, speaking about the ultimate fulfilment of God's plan, who said something like 'I've read to the end of the book—and we win! We win!'

14

THE RISE OF CIVILISATIONS

The First Civilisations

As the inhabitants of the ancient world developed agriculture and animal husbandry, it became possible to move from a nomadic lifestyle to one that was more settled. By around 4000 BC, small family settlements in Mesopotamia were becoming villages and towns. By around 3500 BC, population growth and increased prosperity demanded greater cooperation and organisation within and between communities and the world's first civilisation began to emerge. Some two hundred years later, the Indus Valley civilisation began to take shape, followed about 150 years later by the Egyptians.

Historians tell us that civilisation is defined in terms of seven major components:

- A stable food supply
- A stable social structure
- An effective system of government
- A religious system
- A highly developed culture

- Technological advances
- A method of effective communication over distance and time.

The availability of a steady source of food meant that, aside from intense periods of land preparation, planting and harvesting, inhabitants of towns had a lot of spare time to indulge in a wider range of activities and allowed some members of society to specialise in various crafts. In turn, this led to a society wherein different groups of people were accorded different levels of respect or privilege based on their perceived value to the society as a whole.

Religion played an important role in the development of a civilisation; it helped to define the part that the civilisation itself played within the scheme of things. A characteristic of most of the religions of the times was that both the nations themselves and the multiplicity of gods that they worshipped were in a cooperative relationship—each group to some extent depending on the other. In Mesopotamia, the small group of believers in a single Creator God must have been very much in the minority.

Forms of government were likely to have varied. Earlier forms of government were probably based on single leaders, probably the most senior person in the family or tribe. As numbers grew, there would have been many more people competing for the top position and the 'king' would be the one who was selected by general agreement or by force of arms of the most powerful sector of the society. Initially, many towns were ruled by the priests—after all, who could be better placed to know the will of the gods to which they owed allegiance? However, as time went by, city-states were increasingly led by 'kings' selected because of their perceived ability to defend the city from opposing city-states. The earliest discovered record of a war being fought in Mesopotamia is between Sumer and Elam in 3200 BC.

Another characteristic of civilisations was the culture of the people; most of the early civilisations placed great store by their 'creative' people.

Early Mesopotamian creativity tended to centre on sculptures and building works, particularly in the building of large ziggurats such as the fabled *Tower of Babel*. The earliest known city whose streets were planned and laid out before it was ever built is Mari (built between 3000 and 2900 BC). Significant technological achievements in Mesopotamia during that age were: the invention of the wheel, the development of reliable methods of water delivery and irrigation, writing and sailing boats.

In Mesopotamia, by the middle of the 3rd Millennium, large city states existed, among which were Eridu, Uruk, Ur, Lagash, Kish, Larsa, Mari and Nippur.

The Table of Nations (Genesis Chapter 10)

For the peoples in Mesopotamia during the middle of the 3rd Millennium BC the various flood myths from earlier times, including the account of Noah, were interpreted as having wiped out the entire population of the world (at least the world with which they were familiar). Chapter 10 of *Genesis* is one attempt to account for the spread of humankind after the Flood, starting from just four married couples, one couple of which was well past the age of child-bearing. The other flood myths needed to come up with similar explanations.

Putting aside the whole issue of in-breeding, it is unthinkable that population growth from such a small base could have been sufficient to give rise to so many national groups in what could at best have been only a few hundred years. The best that can be said is that many of Noah's descendants married into and ultimately identified with at least some of these national groups.

In the Genesis 10 account, much is made of *Nimrod*, reputed to be both a mighty warrior and a mighty hunter. He is said to have ruled in *Babylon*,

Erech, *Akkad* and *Calneh* in the region of *Shinar*. It is also suggested that he went on later to *Assyria*, where he built *Nineveh*, *Rehoboth*, *Calah* and *Resen*. The name *Nimrod* is unknown outside of *Genesis* and certainly can't be associated with any of the 'kings' in the Sumerian *King List*.

Akkad was destroyed around 2154 BC. *Erech* (Babylonian *Uruk*) lost its importance around 2000 BC. Therefore, if Nimrod was a real person, he would have had to live some time prior to 2200 BC. Some have suggested that Nimrod might have been the Hebrew name for Sargon (about 2270-2215 BC), who founded the first empire, that of Akkad. There is no evidence to support this.

The Tower of Babel (Genesis 11:1-9)

The story of the Tower of Babel apparently predates the spread of Noah's descendants in Genesis 10 and probably is offered as an explanation for the dispersion of the nations in their own language groups covered in that chapter. *Genesis* does not identify who came up with the idea of building the tower (Babylonian: *ziggurat*) but it was clearly intended to be a joint project. Nor are we told where the tower was to be built.

Remains of at least four such ziggurats can still be found—in *Eridu*, *Ur*, *Uruk* and *Nippur*. All of these were completed. A fifth one probably existed in Babylon but was replaced, at least in part, by Nebuchadnezzar II in the 6th Century BC. Some have suggested that the tower in Genesis 11 is the Babylonian one because of the similarity of the location names (the original Akkadian name for Babylon was *Bab-ilim*, meaning *Gate of God*) but the *Genesis* account says the location of the tower was given the name Babel after the event because it was taken from the Hebrew word *balal*, meaning *confused*. The same passage indicates that both the tower and the city were never completed.

The reason given in *Genesis* for God's opposition to the building of the tower was that men, speaking a single language and working in unity, were becoming very prideful in their ambitions. There is a Sumerian myth called *Enmerkar and the Lord of Aratta* that has an interesting counterpoint to the *Babel* story: *Enmerkar*, the builder and ruler of *Uruk* sets out to build a ziggurat in *Eridu* and demands a tribute of precious materials from a place called *Aratta* for its construction. In the letter of demand, *Enmerkar* offers a prayer to the god Enki to restore the whole territory to use of a single language so that all might address the god in the same language.

The *Genesis* story offers an interesting insight into Mesopotamian building methods that must have been of interest to Moses, educated as he was in Egyptian sciences and technologies. In Egypt, most important buildings were constructed of stone; only ordinary dwellings and public buildings were constructed of mud brick. Mesopotamia, on the other hand, had little access to stone except in the far north and used mud bricks for virtually every kind of building.

The earliest bricks were simply shaped from a mixture of loam, mud, sand and water and were sun-dried until they were strong enough for use. The oldest bricks discovered to date were made before 7500 BC and come from the upper Tigris region. Builders soon learned to mix the mud with a binding material such as straw to increase their strength. From around 4000 BC, all bricks were fire-baked to further increase strength and durability.

The first mortars were made of mud and clay. Later Babylonian constructions used lime or pitch for mortar because pitch was easily found in the area and provided a much stronger bond.

Where is Genesis Taking Us?

My book has been written in an attempt to answer the questions:

- 'How much reliance can we place on Genesis as a set of factual histories?' and
- 'What does the book of *Genesis* teach us about God and his plan for us?'

Up to this point, we have been introduced to God as Creator of the universe and of life. God's relationship with humanity has been limited to an initial revelation of himself to one family. That didn't go well; instead of enjoying the benefits of a God who undertook to provide for and protect them, they chose the ability to make their own decisions about the way they wanted to live. As a result, we have seen God executing judgement—not by destroying them, but in allowing them their independence and the pain and hardships that that entails, and by excluding them from the close personal relationship that they had enjoyed with him. It is possible that God actually did reveal himself to others but with results even more disappointing. For whatever reason, God seems to have placed all of his hopes in a single human family line.

Therefore, for the most part, chapters five to eleven of *Genesis* concentrate on tracing Adam's lineage, of importance to the Jews but of limited value in our search to understand God. Through multiple generations, children were raised on stories of their ancestors' dealings with a single Creator God. Now, they found themselves living in populous, complex and organised civilised societies that worshipped multiple gods. It is probable that many would have had only a nominal belief in one God, but a God who was largely irrelevant to their daily lives; they may have forsaken God altogether (pretty much the way our modern society has done). They seem to have joined with their neighbours in wholehearted general depravity and ambitious pride.

So, when God decided to destroy the other remnants of Adam's line because they had become so thoroughly identified with the violent, depraved society in which they were living, he preserved faithful Noah and his family.

God knows what he is doing. If the genealogies are complete and correct, nineteen generations after Adam God finds someone who wholly believes in the God of his ancestors and is willing to accept God as his absolute lord. This is someone that God can use to put his ultimate plan into action.

Hereafter, Genesis is concerned solely with a single family line beginning with that man and what that family line needed to learn about God in order that God's purposes are worked out. Even so, there is still much that the sciences can contribute to our understanding of that story.

15

ABRAHAM—THE FRIEND OF GOD

Historical Period and Place of His Birth

It is hard to place the story of Abraham within a specific historical period. Outside of *Genesis* there are no corroborative sources to verify that Abraham actually existed. Some modernist scholars have suggested that he was a composite of a number of mythical characters. These same scholars believe that *Genesis* and perhaps the whole Torah was actually written (rather than compiled or copied) by priests during the captivity of the southern kingdom in Babylon because the stories reflected the issues of that time, rather than those of a much earlier period.

While there is little doubt that some editing of Hebrew manuscripts of the Torah was done at that time, the bulk of archaeological and textual evidence suggests that the main content of the original texts remained unchanged. In the past century, archaeological discoveries of documents and monuments from the second and third millennia BC have given us a much greater understanding of ancient Mesopotamia. In particular, excavation of the royal palace at Mari, one of the city states, uncovered thousands of clay tablets of official archives, correspondence and other texts. Many modern scholars claim that these show that the *Genesis* accounts fit perfectly with the social mores, laws and concerns of the early 2nd Millennium BC but only imperfectly with later periods.

Based on studies that attempt to work back from well-dated later events, it is probably safe to say that Abraham lived sometime during the period 2200-1750 BC. Domesticated horses were first introduced into Mesopotamia about 2000 BC; horses are not mentioned in any of the stories about Abraham, which might strengthen the argument for the earlier half of this period.

Abram, as he was first named, was the son of Terah, a descendant of Noah. *Abram* was a popular name throughout Mesopotamia, especially in the early 2nd Millennium BC. The name literally means *father exalted*, which could imply either that the father of the child is thrilled with the birth of a son or that the birth of the son has established the manhood of the father. *Genesis* does state that Abram was born in or around the city of Ur.

Ur was first settled as a village around 4000 BC. Initially, it was located close to the mouth of the Euphrates River and became a great trading centre during the 3rd Millennium BC. Today, as a result of continued flooding and silting of the Euphrates and Tigris Rivers, the ruins of the city now lie some 250 kilometres inland from the Persian Gulf and 10 kilometres from the present course of the Euphrates.

The Ziggurat of Ur probably began construction under the orders of King Ur-Nammu, the first ruler of the Third Dynasty of Ur and completed during the reign of his son Shulgi. The partly-restored remains of the Ziggurat of Ur can still be visited today.

Terah and his sons, Abram, Nahor and Haran, probably lived in Ur in the 21st Century during the Third Dynasty of Ur. This could be considered the 'Golden Age' of Ur. While it used to be thought that the oldest known set of codified laws was the Code of Hammurabi (written about 1772 BC), the more recently discovered Code of Ur-Nammu, was inscribed about

2100—2050 BC. His son Shulgi was responsible for turning Ur into a highly centralized and organized bureaucratic state. As a result, Ur became the centre of culture and learning in southern Mesopotamia. Increasing trade, efficient civic administration and encouragement of artistic and scientific pursuits brought enormous prosperity to the city state.

God states through Joshua (in Joshua 24:2) that Terah 'served other gods' but it seems that Abram held to a firm belief in the one true God that his ancestors told about. We don't know what the family's trade was but a *midrash* (a commentary on Talmudic texts by ancient Judaic scholars) suggests that Abram worked in Terah's idol-making shop, but there is no other support for this. *Genesis* tells us that Haran died in Ur, leaving a son, Lot. It is probable that Terah's family, living in a prosperous community, were quite well off.

Despite Terah's evident prosperity and the prestige of living in Ur, we are told in Genesis 11:31 that Terah took Abram, Sarai (Abram's wife) and Lot and set out for Canaan. We are not given a reason but there are at least three possibilities:

- He might have been trying to develop a trading business between Ur and Canaan. He does not take Nahor, who may have been left behind to manage the Ur end of the business;
- For about 700 years, 2600–1900 BC, there was intense rivalry and animosity between the city states of Southern Mesopotamia, with almost continuous struggles between them for control of the whole region. It could well be that Canaan appeared to Terah to offer a more settled and peaceful place to live;
- In Acts 7:2, Stephen in his address to the Sanhedrin states that Abram was called by God before he went to Haran, so it is possible

that Abram was the one who prevailed upon his father to leave Ur. God does not appear at that time to have told Abram *where* to go.

Given that Abram's antecedents were by and large a sorry lot in terms of their faithfulness to their God, Abram needed to be thoroughly separated from his past ties if he was to enter into a close relationship with God. Thus it was essential for him to leave his country, his culture, his acquaintances and his father's household (remember that Terah was a worshipper of idols—every city state had its own gods and every home had its own household gods).

Regardless of the reason, they left Ur and travelled through the Fertile Crescent until they reached Haran, where they stopped. Again, we don't know why, but the region was certainly more peaceful and apparently provided attractive business opportunities. If the move was related to God's call to Abram, it is clear that Abram failed to separate from his father's house, either because he could not bring himself to do so or because Terah himself may have decided that the move was good for the whole family.

The Move to Canaan

It is not clear whether God's words to Abram in Genesis 12:1-3 are the original promise made to Abram in Ur or are a renewal of the promise while they were residing in Haran. Nevertheless, there is now no doubt that God's plan to establish a relationship with his creation is about to move into its next phase. We can see this unfolding in the six parts to the promise:

- 'I will make you into a great nation'—not a great man or a great ruler; but Abram's offspring will become a great nation;
- 'I will bless you'—can be taken to include safety, health and wealth;
- 'I will make your name great'—someone of importance and influence, someone to be remembered with respect and affection;

- 'you will be a blessing'—someone who will bring benefits to others through association with him;
- 'I will bless those who bless you and whoever curses you, I will curse'—God will take Abram's part in every situation;
- 'all peoples on earth will be blessed through you'—as the vehicle for the fulfilment of God's plan to make himself known to his creation.

We are not told how God 'spoke' to Abram. The most likely way was in a dream or a vision as these were highly regarded in those times. (It is worth noting that these are still frequent occurrences in this day and age with spectacular results, as close personal friends of mine have experienced.)

We also don't know how long Abram stayed in Haran, only that he was 75 years old when he left Haran. He did not, as some have suggested, wait until Terah died; Terah had turned seventy when Abram was born and so was about 145 at this time. Terah lived to be 205. We learn in *Genesis* chapter 24 that Abram's surviving brother, Nahor, eventually moved to the area around Haran, establishing his own 'town'; it is likely that Terah had been left in his care. This is further supported by Genesis 29:4-6 where Nahor's grandson lives in Haran.

Abram had obviously prospered in Haran because his caravan includes much livestock and a large number of servants and slaves to care for them. Abram's decision to take Lot to Canaan with him did not mean that he was being disobedient to God's instruction to leave his father's house; as Abram was the older brother of Haran, Lot's dead father, he effectively became responsible for Lot, who became part of his household.

Abram's first major stop in Canaan was at Shechem, which was a major centre of Canaanite Baal worship. The purpose of this stop was probably to make the Canaanite leaders aware of his presence in their land, but God 'appeared' to him there (whether in a dream, a vision or in human likeness we are not told) and promised that this land will be given to Abram's offspring (of which Abram had none at the time). Abram's response is to build an altar to God in that place. We can only speculate about the Canaanite's reaction to someone building an altar to another god in an area that was a centre of Baal worship!

Abram moves on to a location that will later become known as *Bethel*, meaning *House of God*. It is clear that this area was regarded by Abram as being a place where he wanted to settle; here he builds a second altar and calls on the name of the Lord (*Yehovah*) before continuing his exploration of the land in the direction of the Negev (the dry wasteland south of Beersheba).

Abram in Egypt

A famine in Canaan drove Abram south to Egypt where the Nile River could generally be relied upon to water crops, even in time of famine. This is assumed to be quite soon after Abram entered Canaan because he had not yet settled. We cannot assign a more accurate date for this move to Egypt except that it almost certainly occurred during the time of Egypt's 'Middle Kingdom' (the dates of which have variously defined as 2060-1802 BC and 2040-1782 BC). This was the 'golden age' of Egypt, with the finest art and literature to ever be created in the long history of Egypt.

On their journey, Abram was concerned that his wife Sarai was so beautiful he feared that the Egyptians would be so taken with her that they would kill him to possess her. He asked her to pose as his sister. Now Sarai had to be at least 65 at this time, so how realistic were Abram's fears?

We know that standards of beauty vary widely from society to society and age to age. We can't be sure what the Egyptian idea of beauty was at the time, but one characteristic was known to be highly prized—fairness of complexion—and this was certainly something that all Mesopotamian women had. Taken together with a high degree of sophistication learned in Ur's social milieu, Sarai was likely to be a striking figure.

Here we learn something new about God and about Abram:

- God did not choose Abram because he was perfect—only that he was sincere in his belief in God and willing to go where God specifically led him; and
- Even having received God's promises, Abram still depended on his own misguided wisdom to protect himself (in spite of the fact that he was putting Sarai at risk of being taken from him).

Sarai did indeed attract a lot of very favourable attention and was taken into Pharaoh's palace, presumably to become a concubine; Abram did very well out of the deal, being given in return sheep, cattle, donkeys, male and female servants and camels.

Some scholars have jumped on this mention of camels as clear evidence that *Genesis* was written at a much later date because no camel remains have been discovered in Palestine dating any earlier than nearly a thousand years after this time. However, some years back a *petroglyph* (picture made by chipping into a rock surface) of a camel was found in the Negev that dates from before 5300 BC, and it has recently been established that camels may have been domesticated in Somalia and southern Arabia as early as 3000 BC. As domesticated camels were likely to have been very rare at this time and highly prized, lack of discovered remains does not entirely rule out the possibility of their existence. It is even possible that Abram, migrating from Mesopotamia, had the only camels in Canaan at the time.

God had not been party to the deal concerning the fate of Sarai and once more intervened in man's affairs, bringing various sicknesses into the palace. Pharaoh's wise men were quick to pinpoint the source of the problem ... Sarai was already married.

One solution for the Pharaoh would have been to do what Abram had feared and kill Abram or even kill both Abram and Sarai. Whatever might have motivated Pharaoh, God was at work and Abram and his people were sent packing ... together with all that Abram had received from the Egyptians.

Return to Canaan

As I have said earlier, this book is not intended as a commentary on *Genesis* except in relation to what light science and archaeology can throw on it, so I will pass over much of this *Genesis* story. However, there are a number of incidents and details that are worthy of mention.

Abram and Lot returned to Canaan as very rich men and by virtue of their wealth, numbers of servants and size of their flocks, exerted a great deal of influence in that land. They soon found themselves faced with the problem of the size of their respective flocks and herd which not only put them in competition with one another for grazing land, but tended to dominate the arable land and pastures used by the Canaanites. Their solution was to separate and, given the choice by Abram, Lot decided to move down into the lowland plains around the Jordan River and the Dead Sea and eventually settled in Sodom—the major town in the area—exchanging his tent for a house. There is a very interesting comment within the *Genesis* story:

> *Lot looked around and saw that the whole plain of the Jordan toward Zoar was well watered, like the garden of the LORD, like the land of Egypt. (This was before the LORD destroyed Sodom and Gomorrah.) (Gen 13:10)*

As mentioned earlier, there are some Bible scholars who maintain that *Genesis* was written in Babylon during the Judean people's captivity under Nebuchadnezzar. For them, this verse constitutes a problem. I am not the first to point out that, if this had been written by Judean priests in Babylon, they would have no knowledge of the area of Egypt referred to, and it would be meaningless to the captives. But if Moses was the author or supervised its writing for a people that he was in the process of leading out of Egypt, they would have been very familiar with the area of Egypt that he described. The passage describes the Jordan Valley in glowing terms as a 'garden' in the days of Abram and Lot. That clearly was not the way that the Children of Israel saw it as they entered it the first time to send out spies to explore Canaan. At that time, the southern end of the valley in particular was poorly watered and sparsely populated. So Moses had to remind the Children of Israel that the destruction of Sodom and Gomorrah had had very adverse consequences to the landscape through which they travelled. If this had been written during the Babylonian captivity, such an explanation would have been unnecessary. But we will return to the geography of Sodom and Gomorrah shortly.

When we think of armies and battles at the time, we need to understand that these conflicts were on a very small scale. Each of the large towns on either side of the Jordan and around the Dead Sea had their own 'king'. *Kedorlaomer*, the 'king' of Elam, with the help of four 'kings' from nearby towns set out to conquer and control the general area. After conquering a number of towns, they were eventually opposed by an alliance of five kings, including those of Sodom and Gomorrah. Kedorlaomer's army defeated the five-king alliance and proceeded to loot Sodom and Gomorrah, carrying off Lot and his household. A survivor managed to make it back to find '*Abram the Hebrew*' (Gen 14:13). This is the first time that *Hebrew* is used in *Genesis* and marks Abram as the 'father of the Hebrews'.

When Abram was informed, he immediately gathered the 318 trained men of his own household, (apparently augmented by a small number of men supplied by Canaanite leaders Aner, Eshcol and Mamre) and set off in pursuit. While it is probable that Kedorlaomer's army had suffered some losses in his previous battles, Moses small army was quire sufficient to route the raiders; Abram was able to free all of the captives and recover all of the goods looted from the two towns.

It is interesting to note that the story refers to the Valley of Siddim, where the battle between the four kings and the five kings took place, as being 'full of tar pits'. We will return to that when we consider the fate of Sodom and Gomorrah.

Establishing the Covenant

In Genesis 15, we are told of Abram's next direct revelation from God, this time in a vision. Up to this point, Abram had shown himself to be obedient in the things he was asked to do. In the vision, God re-committed himself to making of Abram a great nation. Even though Abram was quick to point out that he was childless, God assured Abram that he will have a genuine son of his own and told Abram what was going to happen through that child over the next several hundred years to the time when his descendants would take possession of the whole of the country in which Abram was currently residing. For God's part in establishing the covenant, he symbolically passed between the parts of animals that Abram had killed and cut into halves. This is believed to have been a form of binding contract of the time, in which the person passing between the parts of the animal prepared by the other person is committing themselves on oath that effectively says *'May it be so done to me if I do not fulfil my pledge'*.

At this stage God was making a very one-sided commitment to Abram and it was probably some twenty years later when God required Abram to commit himself and his descendants to their side of the covenant through the practice of circumcision.

When Abram and Sarai had lived in Canaan for ten years—at which time Abram was eighty five and Sarai was seventy five—and Sarai was still not pregnant—they thought it was time to help God out. In keeping with accepted practice of the time, Sarai gave her young Egyptian maidservant Hagar to Abram so that he might have children by her. There is no reason to assume that this was in any way part of God's plan, but God graciously blessed Hagar and her son Ishmael, who founded his own dynasty under God's permissive hand. Today, the Arabs regard Ishmael as the first-born and rightful heir of Abraham and their 'father'; thus their hostility to the Jews (not discounting their anger at the dispossession of their homes and land by the Jews in 1948).

A number of people groups around the world have long practiced circumcision as a rite marking the passage of a boy into manhood. Arabs circumcise boys at the age of thirteen as both a mark of entry into manhood and because Ishmael was circumcised at that age. Circumcision as decreed by God for Abram's descendants is different in both its significance and practice. To the Children of Israel, circumcision signifies their commitment to and participation in God's covenant with Abram, whom God renamed as *Abraham* (meaning *father of many*). Rather than a rite of entrance to manhood, Abrahamic circumcision is symbolic of an oath to the effect that *'If I am not loyal in faith and obedience to the Lord, may the sword of the Lord cut off me and my offspring as my foreskin has been cut off'*. Circumcision was to be performed at the time a male became a part of the household of Abraham (even as a servant or slave); new-born males were to be circumcised on the eighth day after birth.

Intervention and Intercession

In the stories of the Flood and the Tower of Babel, we have seen that God can choose to intervene in this world where groups of people become so disruptive in their behaviour that whole young civilisations could be damaged irrevocably. In Chapter 13, I have maintained that God is not caught off guard by man's exercise of his free will and is able to intervene at any point in time to ensure that his plan accommodates man's actions. The story of Ishmael is a good example. But Chapter 18 of *Genesis* teaches us something new in the relationship between God and mankind.

Chapter 18 is the first time that *Genesis* is unequivocal in stating that God appeared before Abraham in human form, this time in the company of two angels (also in human form). The visit is witnessed by Sarah and the servants. Abraham's hospitality was an accepted Middle Eastern reception for any strangers who may be of some account, so it is unclear whether Abraham immediately recognised the true nature of the three visitors or only became aware of this when God referred to Abraham's wife as Sarah, the name that God had given Sarai at the time when the covenant of circumcision was instituted. What is clear is that all three of the heavenly visitors accepted and partook of Abraham's hospitality.

Afterwards, Abraham accompanied the three on their journey for a respectable distance. During this journey, God told Abraham of his decision to destroy the cities of Sodom and Gomorrah because of the sins of the men of those cities were so blatant that even the inhabitants of the surrounding regions were appalled. The actual nature of those sins is not stated explicitly, but the description of the later behaviour of the men of Sodom towards Lot and the angels supports the suggestion that the sins were probably ones of sexual depravity, including both heterosexual and homosexual rape.

In revealing his intentions to Abraham, God effectively made Abraham his counsellor and was asking for his take on the situation. In one respect, this was a test for Abraham—was he a 'yes man', prepared to go along with anything God suggests in order to stay on God's good side? Or was he a man of integrity who would stand up for what he believed, even if this required him to disagree with the Creator of the Universe?

Regardless of its merits as a test for Abraham, the most amazing thing about the situation is that God actively encouraged a human to influence God's plans. A common characteristic of most religions that are centred on a belief in one or more gods is that in some way the gods must be cajoled or bribed into showing favour to believers. Commonly, this involved human sacrifice or self-mutilation to impress the gods with how serious they were in wanting an answer. But here God shows us that he encourages our willing participation with him as co-workers with him in the realisation of his plans for mankind. This is the basis on which we begin to understand the nature and privilege of intercessory prayer.

Abraham's main concern was for God's reputation and so asked God to do what a righteous judge would do and spare the righteous, even if that meant that the sinners go unpunished. Abraham asked God to spare the cities if fifty 'righteous' people can be found there. Having God's assurance that in that instance the cities would be spared, Abraham gradually argued for progressively smaller numbers to be accepted as the basis for sparing the cities and God finally agreed to spare the cities if only ten 'righteous' persons were found.

God is no fool; he is not being 'squeezed' into a corner. He is showing Abraham (and us) that he not only can change his mind, he is willing to change his mind in response to our prayers, provided that it can be accommodated within his overall plan.

Sodom and Gomorrah

In the Western World, even people who have little Bible knowledge have heard something of Sodom and Gomorrah, if only as cities that had been destroyed by God for their wickedness. But, to date, archaeologists have not been able to clearly identify any remains that provide incontrovertible proof that they ever existed, despite many decades of research. It is appropriate that we spend some time on the subject.

Location

The Greek geographer and historian Strabo, writing at the time Jesus Christ was growing up in Nazareth, was told by locals living around the southern end of the Dead Sea that there were once thirteen inhabited cities in that region of which Sodom was the principal city. He identified a limestone and salt hill at the south western tip of the Dead Sea and ruins nearby as the site of biblical Sodom. The hill still bears the name *Jebel Usdum* (*Usdum* being another variant of the word *Sodom*). During the second half of the 20th Century, a number of other sites have been investigated both northeast and southwest of the Dead Sea. While the majority of contemporary scholars still favour the region southwest of the Dead Sea, some feel that the northeast region better suits the Biblical description of an area that was well-watered and fertile. Of course, that description referred to a time well before the massive destruction of the two cities and we have already noted that the Children of Israel found an area that had changed considerably from the time of Abraham.

Apart from the writings of Strabo, evidence for the region southwest of the Dead Sea includes the references to the tar pits in the Vale of Siddim (Gen. 14:10), a feature of the southern end of the Dead Sea, and to Zoar, a township on slightly elevated ground in what later became Moabite territory (southeast of the Dead Sea). From the description of Lot's flight from Sodom (Gen. 19:17-22), Sodom could not have been far from Zoar.

How Were Sodom and Gomorrah destroyed?

The Biblical account of the destruction in Genesis 19:23-28 reads:

> *[23] The sun had risen on the earth when Lot came to Zoar.*
>
> *[24] Then the LORD rained on Sodom and Gomorrah sulphur and fire from the LORD out of heaven.*
>
> *[25] And he overthrew those cities, and all the valley, and all the inhabitants of the cities, and what grew on the ground.*
>
> *[26] But Lot's wife, behind him, looked back, and she became a pillar of salt.*
>
> *[27] And Abraham went early in the morning to the place where he had stood before the LORD.*
>
> *[28] And he looked down toward Sodom and Gomorrah and toward all the land of the valley, and he looked and, behold, the smoke of the land went up like the smoke of a furnace.*

Sceptics of the biblical account suggest that, if the cities existed at all, they were probably destroyed by natural disaster. I agree with them. I see no reason to believe that God created a storm of burning sulphur out of nothing; on the contrary, I believe that God usually achieves his purposes by entirely natural means where this is possible. For instance, in Joshua 3:14-17 we are told that the Jordan, at its greatest annual flow, was stopped by what is almost certainly a landslide some distance up river precisely at the right time so that the Children of Israel could pass over on dry ground; when all of the people had crossed, the river overflowed the stoppage and resumed its normal flow. A note appears against Joshua 3:13 in the *New International Version Study Bible (1984)* states that: 'As recently as 1927 a blockage of the water in this area was recorded that lasted over 20 hours.'

So how might the destruction of Sodom and Gomorrah have been achieved?

It is possible that an earthquake released high temperature tars and other volcanic materials that had been trapped under high pressure into the atmosphere to rain down on the cities. While this is theoretically possible, the *Genesis* account does not mention any shaking of the ground—only on material raining from the skies.

There are ten recently active volcanic fields in the Arabian Desert region (that is, active within the last 10,000 years). Six of these have erupted within the last 2000 years and two within the last 200 years. This gives rise to another possibility - that an eruption of one of these volcanoes at that time produced a rain of volcanic material that fell across the desert areas to the northwest as far as the Dead Sea end of the Jordan Rift Valley. Such an event could certainly produce the destruction that *Genesis* describes, but I can find no mention of any research that might support this possibility.

A third possibility is that these cities could have been destroyed by the explosion of a meteorite or comet entering earth's atmosphere above the plain where they were situated. There are historical records of such events:

- In 1908, a comet or meteor blast over a sparsely populated region of Siberia, known as the Tunguska event, flattened 2,000 square kilometres of forest, fortunately without loss of human life.
- In 2013, a meteor explosion over Chelyabinsk in Russia injured more than 1,600 people, mainly due to broken glass from windows shattered by the blast.

Recent news published by members of the *Tall el-Hammam Excavation Project* has brought this possibility more centre stage. Since 2005, this group from Trinity Southwest University in Albuquerque, New Mexico and the Veritas International University in Santa Ana, California has been engaged in the long-term excavation of *Tall el-Kammam*, an area just northeast of the Dead Sea in Jordan. The driving force behind this project is a belief that this site is the real location of Sodom, despite most mid-eastern archaeologists having rejected the idea.

In 2018, the project team published the results of research that indicated that a superheated blast from the skies obliterated cities and farming settlements in a 25-kilometer-wide circular plain called *Middle Ghor* on which Tall el-Hammam is situated. What had been a densely populated and fertile area was suddenly devastated; radiocarbon dating indicates that the mud-brick walls of settlements suddenly disappeared about 1700 BC, leaving only stone foundations. The fertility of the land was destroyed and the whole area remained sparsely settled for around 700 years.

The team suggested that the event was the result of a meteorite exploding at a low altitude with a force equivalent to 10 million tons of TNT. The fireball created tiny, spherical mineral grains that rained down on Tall el-Hammam; the research team has identified these minuscule bits of rock on pottery fragments at the Tall el-Hammam site. In addition, outer layers of many pieces of pottery from same time period show signs of having melted into glass. Zircon crystals in those glassy coats can form only at extremely high temperatures. The event splashed hot Dead Sea brine across what had been fertile land.

If this interpretation of the evidence is correct and the angle and direction of the meteorite's fall is appropriate, it is highly possible that the blast could have sent a tsunami of boiling brine down the length of the Dead Sea to inundate the plain to the sea's south; this would certainly be sufficient to destroy the cities of the plain. This would also explain the *Genesis* account of Lot's wife being overwhelmed by and becoming encased in salt.

At present, this theory is almost entirely speculative and needs a lot more investigation before we can place much reliance on it. Moreover, the dating of the event around 1700 BC is perhaps 400 years too late to fit with the general consensus that Abraham lived in the late 2nd Millennium BC or early 3rd Millennium BC. The later date would move the Exodus from late Second Intermediate Period or early New Kingdom Period of Egyptian history (about 1550 BC) to around the time of Pharaoh Rameses II. (This later date used to be considered to be the most likely date for the Exodus, but recent scholarship favours a much earlier date, as we will see later in this book.)

The best that can be said at this stage is that this theory merits further research.

Why Haven't Sodom's Remains Been Found?

While the remains of ancient settlements are scattered all over the Middle East (some of which are being excavated as possible remains of Sodom and Gomorrah), nothing has been found in the area believed to be the most likely location of these cities—the plain south of the Dead Sea. To say the least, it is unusual for no remains to be found where a sizeable city is known to have existed, which renews the question: 'Did Sodom and Gomorrah ever exist in that area?'

Early historical writers believed that the sites of Sodom and Gomorrah lie beneath the waters of the Dead Sea. This is certainly a possibility and submersion may have occurred at the time of destruction or in the years since. Geological research gives rise to a number of possibilities, not all of which require the remains to lie under the Dead Sea.

In the early years of last century Alfred Wegener, a German polar explorer, geophysicist and meteorologist, noticed that if the continents, including their continental shelves, were cut out of a map of the world, they could be fitted together as in a jigsaw to form one large super continent. In 1912, he proposed a *Theory of Continental Drift* in which he suggested that there had originally been only one very large continent, that this had split into pieces and that the individual pieces had 'drifted' apart. Comparing rock types, geological structures and fossils, he was able to show that there were significant similarities between matching sides of the continents. Unfortunately, he could not explain how or why such drifting might occur and the theory was largely ignored.

It was not until the middle of the century that a better understanding of the structure of our planet allowed such a mechanism to be identified and, over the next quarter-century, to be validated. The mechanism is called Plate Tectonics.

Put simply, Earth consists of four parts—a solid inner core made up largely of iron and nickel at a temperature of up to 5500°C and with a radius of about 1300 km, a 2250 km thick liquid outer core with the same composition, a viscous 'mantle', 2900 km thick and a cool outer crust, the part on which we live. The thickness of this crust varies from place to place, but is generally between 8 and 40 kilometres thick. Within the viscous mantle, hot material near the core rises towards the surface, just as hot water in a saucepan rises, and colder material sinks towards the core. This movement is referred to as *convection* and it is this constant circulation of material in the mantle that causes the crust to break up into 'plates' that move relative to one another according to the direction of the flow in the mantle material below them.

Most volcanoes are found along the boundaries between plates and most earthquakes occur along those same boundaries as the plates 'grind' against each other, pull away from each other or crash into each other, sometimes with one plate sliding under the other. The speed with which these interactions take place varies but probably does not anywhere exceed 10cm per year.

The Jordan Valley and the Gulf of Aqaba are part of the Dead Sea Rift, a great scar formed in the Earth's crust as the Arabian Plate on the east of the rift pushes north into Iran, where it is continuing to build mountains, and the African Plate on the western side is pivoting to the south and west. The relative movement between the two plates has created a deep crack in the crust that widens and lengthens each year. The plates are passing each other at the rate of about 5 mm per year. This movement gives rise to two phenomena that are relevant to the story of Sodom and Gomorrah:

- As the crack lengthens and widens, more of the sedimentary material in the base of the valley sinks into it, which is why the valley bottom at the southern end is now below sea level. The rate of fall of the valley floor isn't constant but mainly occurs during large earthquakes. Averaged over time, it has been suggested that the floor of the valley at the southern end is falling around an inch (2.5cm) a year.
- In recent years the Dead Sea area of the Jordan Valley has experienced an average of about one earthquake measuring over 5 on the Richter scale every four years. When soil is saturated or partially saturated, the shaking of the ground by a strong earthquake can cause the ground to lose its strength and hardness and behave like a liquid such that whole buildings simply sink below the surface. This phenomenon is called *soil liquefaction*.

So then, we have three possible reasons why Sodom and Gomorrah may never be found—the ruins lie beneath the Dead Sea or they have been swallowed up as sedimentary materials fall into the widening rift or they have been the victims of soil liquefaction. Nevertheless, I live in hope that one day we will find enough evidence of their existence to satisfy the most sceptical critics.

The Birth of Isaac

Abraham was one hundred years old when Isaac was born and Sarah was ninety. Today, with all of the advantages of modern medical science, this is considered to be unlikely, if not impossible. In September 2019, an Indian woman of 74 became the oldest person to have given birth (in this case, to twins) and this was achieved through IVF. The father was 82. The oldest woman known to conceive naturally and give birth successfully to date was 71.

Isaac's birth can rightly be classed as miraculous, particularly since it was announced twelve months in advance by God. Everything about this child was to be special because he was the child of the promise made by God 25 years before (Genesis 12:2-4). Why did God wait so long? We don't know, but it may have been to ensure that Abraham's line was synchronized with future events that God would use to further his plans (such as the timing of the Egyptian pharaonic dynasties). However, this is only speculation.

One thing is certain—Abraham, at an age when he was beyond all hope of becoming a father, now had another son and, on God's instruction, a sole heir.

The Testing of Abraham

Genesis 22:1 tells us that some years later, when Isaac was clearly an adolescent, God spoke to Abraham. There is no suggestion that God appeared in any visible human form, or in a vision or dream, but rather as a voice, much as happened initially to Samuel in 1 Samuel 3:1-8. The command God gave to Abraham must have seemed as unreasonable and cruel to Abraham as it seems to us today:

> *After these things God tested Abraham and said to him, "Abraham!" And he said, "Here I am." He said, "Take your son, your only son Isaac, whom you love, and go to the land of Moriah, and offer him there as a burnt offering on one of the mountains of which I shall tell you." (Gen 22:1-2)*

Nothing in his relationship with God could have prepared Abraham for this. The Canaanite gods might have condoned, if not required, human sacrifice, but surely not the Lord God!

There was no doubting Abraham's deep love for and pride in Isaac, especially given that God had told him that it was through Isaac that God's covenant would be fulfilled. But early the next morning he set out with Isaac, two servants, wood and fire to do what the Lord God had asked of him.

We can only imagine what was going through Abraham's mind on the three-day journey from Beersheba to Moriah, but two things that Abraham says on the way offer some idea that he thought God would intervene in some way:

> *On the third day Abraham lifted up his eyes and saw the place from afar. Then Abraham said to his young men, "Stay here with the donkey;* **I and the boy will go** *over there and worship* **and come again** *to you." (Gen 22:4,5)*

> *And Isaac said to his father Abraham, "My father!" And he said, "Here I am, my son." He said, "Behold, the fire and the wood, but where is the lamb for a burnt offering?" Abraham*

> said, "***God will provide for himself the lamb*** *for a burnt offering, my son." So they went both of them together. (Gen 22:7,8)*

But there was no lamb. Abraham finally had to face the fact that God was really asking him to kill his son. He bound Isaac (what was Isaac thinking when this happened?) and it was not until Abraham had tensed to plunge the knife into Isaac's heart that an angel called out to stop him—and only then did Abraham find the ram that God had provided.

What was the point of the whole story? Many sceptics claim that God, if there is a God, is a cruel god that appears to toy with us, that seems indifferent to our suffering and perhaps even enjoys seeing us suffer. Is this just one more instance of God filling us with hope then dashing it, only to give us new hope and starting the whole torture cycle over again?

So, if this story is true and comes from God, what was its purpose and what was it meant to teach us?

Putting God First

Firstly of course, it was significant in the relationship between God and Abraham. It proved to God how committed Abraham was to him … that in fact Abraham was determined to obey him regardless of the cost. But it also proved to Abraham how committed he was to God.

In Luke 14:26, Jesus says: "If anyone comes to me and does not hate his own father and mother and wife and children and brothers and sisters, yes, and even his own life, he cannot be my disciple". Of course, he was using hyperbole to emphasize that we must put God so far above any other relationship that it is almost as if we hated them. The use of hyperbole to emphasize a point was common throughout the world of Jesus' time, not just amongst the Jews, and Jesus used this figure of speech often in his teaching. In Abraham's day, thought and language was far less developed; when God needed to make a point, it had to be very direct.

It is pretty easy these days to convince ourselves that we are totally committed to God, but how would we stack up against Jesus' hyperbolic statement or what Abraham was asked to do? Can we really know how genuine our commitment is until all that is most important to us is threatened?

What would have happened if Abraham had refused God's command or had tried to go through with it but folded before the task was completed? We don't know. God had already made his covenant with Abraham and would not break it—the future of his family was assured. But now, because he had obeyed, he knew for certain that God could be fully trusted and God knew that he could fully trust Abraham. In Isaiah 41:8 God calls Abraham 'my friend'. Friends can rely on each other.

For Christians, this story has very great importance because of its symbolism. Not since God told Adam that one of his descendants would be attacked by Satan but that person would in turn crush Satan's head do we have any clear promise of a saviour to come. Abraham's willingness to sacrifice his own son out of love for God speaks to us of God's willingness to sacrifice his own son out of his love for us. The provision by God of an alternative sacrifice is a clear example of his willingness to provide, and to accept, a substitutionary sacrifice.

Moriah

It is appropriate at this point to address an issue that Jews in particular hold to be very important. Note that Abraham was instructed to go to the *region of Moriah* in order to sacrifice Isaac. In the *Genesis* passage, the Hebrew word for *region* is *erets*, a word we've seen before in the story of Noah where it was translated as *the world* and in fact can be anything from the *whole (known) world* down to *a piece of ground*. Moriah is unknown outside of this passage although it was obviously a mountainous region and had to be reachable on foot from Beersheba within three days, so it was probably located in the Judean high country.

In 2 Chronicles 3:1, we are told that the temple in Jerusalem was built on Mount Moriah and was the place where God had appeared to King David. Some people have related the two passages to suggest that the temple in Jerusalem was thus built on the very spot where Abraham had tried to sacrifice Isaac, but this is almost universally rejected by scholars for two reasons:

- In Genesis 22:2, the area is referred to as Moriah and God says that he will direct Abraham to a mountain within that area. 2 Chronicles 3:1 refers to a mountain called Moriah; and
- At the time of Abraham, Jerusalem already existed as a large enough city to have its own king, *Melchizedek* (Gen 14:18), so it would be unlikely that Abraham would have been sacrificing Isaac there.

We do know that the books of 1 and 2 Chronicles were written in Babylon during the time of the captivity, but we have no idea when (or whether) the temple site was given the name of Mount Moriah. The name does not appear anywhere else in the Bible. The use of the name may have been introduced to relate the temple site to the site of Abraham's test, but even the priests who wrote 2 Chronicles did not make any statement relating the two.

A Note on Ishmael and Islam

Islamic Prophets

The Arab nations regard Ishmael as their father and there is good reason to accept this. All that is known of Ishmael from ancient writings comes from *Genesis*. Muhammad originally was very influenced by Judaism and Christianity and much of Islam is based on the Bible; of the 25 prophets recognised by Islam, only Muhammad himself is not a Biblical character. In Islam, a 'prophet' is not necessarily a bringer of a spoken or written message from God; a prophet might only be someone sent by God to be an example of how a true follower of God should live. The recognised prophets are (Islamic name followed by the Biblical name):

ʾĀdam (Adam), ʾIdrīs (Enoch), Nūḥ (Noah), Hūd (Eber), Ṣāliḥ (Salah), ʾIbrāhīm (Abraham), Lūṭ (Lot), ʾIsmāʿīl (Ishmael), ʾIsḥāq (Isaac), Yaʿqūb (Jacob), Yūsuf (Joseph), Ayūb (Job), Dhul-Kifl (Ezekiel), Shuʿayb (Jethro), Mūsā (Moses), Hārūn (Aaron), Dāūd (David), Sulaymān (Solomon), Yūnus (Jonah), ʾIlyās (Elijah), Alyasaʿ (Elisha), Zakarīya (Zechariah), Yaḥyā (John the Baptist), ʿĪsā (Jesus), Muḥammad.

Abraham and Ishmael

In the Qurʾan, Muhammad states that Abraham was asked by God to sacrifice his son, but the name of the son is not given. This was a test and, before Abraham can actually kill his son, God stops him. Unlike the Bible, no mention is made in the Qurʾan of God providing an animal as an alternative sacrifice. Most Muslims believe that the son being sacrificed was Ishmael but there are no pre-Qurʾanic documents that support this belief.

16

ISAAC–CHILD OF THE PROMISE

The Two Brothers

When Abram was 75 years old and childless, God made him a promise that he would have a child ('seed') to whom God would give, as a possession, all of the territory in which Abram was living as an alien. When Abram was 85 years old, he was still childless. Ten years and God had still not fulfilled that promise. But we are told that God once more appeared to him in a vision and renewed that promise and '*he* [Abram] *believed the LORD, and he* [God] *counted it to him as righteousness.* (Gen 15:6)

Once God had gone to so much trouble to confirm his promise (described in Chapter 15 of *Genesis*), Abram and Sarai made the mistake of assuming that God needed their help, so they took things into their own hands. At 75 years of age, Sarai knew she was too old ever to have children and gave her young Egyptian handmaid, Hagar, to Abram so that he could have a child through her. And so Ishmael was born.

Ishmael was not part of God's plan but, God was gracious enough to still care for Hagar and Ishmael and they continued to be part of Abram's family for the next fifteen years. As his first-born son, Ishmael was greatly loved by Abram.

When Abram was 99 years old, a short time before Sodom and Gomorrah were destroyed, God once more appeared to Abram and announced that the promised son would be born of Sarai about a year hence. It was on this occasion that God gave Abram the new name of Abraham ('Father of Many') and Sarai the new name of Sarah.

Even after God assured him that Sarah will at last give birth to the long-promised child, Abraham still wanted Ishmael to be the inheritor of the covenant. Abraham had wanted to solve God's problem with his own effort and produced Ishmael. But God makes it clear that his covenant with Abraham will only be continued through the child of the promise—to be named Isaac. In effect, God's response showed that his purposes are not fulfilled through our efforts but by God's power. The importance to us of this response is stated in Paul's Letter to the Romans:

> *But it is not as though the word of God has failed. For not all who are descended from Israel belong to Israel, and not all are children of Abraham because they are his offspring, but "Through Isaac shall your offspring be named." This means that it is not the children of the flesh who are the children of God, but the children of the promise are counted as offspring. (Romans 9:6-8)*

Here he is making the point that genetic descent from Abraham or Israel is not what makes a person acceptable to God, but the true 'children of God' are those who come to God through the way that god has specified—that is, through repentance and faith in Jesus Christ.

Despite the fact that Ishmael could be regarded as 'surplus to requirements' and irrelevant to his plan, God still honours Abraham's love for Ishmael and graciously promises that Ishmael too would become the father of a great nation.

Isaac and Rebekah

There is little of scientific interest in the tale of Isaac and Rebekah as told in Chapter 24 of *Genesis*, but we are given a wonderful insight into the nature and power of prayer.

When Isaac became an adult, it was necessary for him to marry in order to maintain the line. Abraham recognised the dangers of Isaac marrying a Canaanite woman, since this would involve developing unacceptable legal relationships with the Canaanites and risk Isaac being pressured into recognizing the Canaanite gods. So Abraham commissions his chief servant to go to the area around Haran to find a suitable bride from amongst Abraham's own family.

We aren't told the servant's name; it may have been Eliezer of Damascus who had been Abram's chief servant and heir some fifty five years earlier, before Abram's first son was born. As a servant, his own religious faith was probably based on that of his master, but without the genetic heritage or the deep personal relationship with God that Abraham had over many years.

The servant finds his way to the general area where Abraham's relatives lived and then, in an act of faith towards a God he did not personally know, prays 'O Lord, God of my master Abraham', followed by a very detailed prayer that he might be guided to the woman that God had chosen for Isaac. And God answers exactly in accordance with that prayer. So thrilled and overwhelmed is the servant for this answer to prayer that he recounts the prayer in great detail to Rebekah's family as evidence that Rebekah is God's choice for Isaac.

This is the first time in the Bible that we learn that it doesn't require great knowledge about God or a deep personal experience with God for us to be able to pray to him in difficult times and have him answer. This lesson is one that is found many times throughout the Bible and one that thousands of people down through the last 2000 years have found to be true.

Jacob and Esau

Isaac and Rebekah only had two children—twin boys. Before they were born, God told their mother that both boys would found new nations but that the elder would serve the younger. This should not be taken as a case of God arbitrarily choosing one over the other but simply as a statement of fact because, as we saw back in Chapter 13 of this book, God knew which of the boys would eventually commit himself to God.

From the first, the twins could hardly have been more dissimilar. Esau (*'Hairy'*), the eldest, was red-headed and hairy, an outdoors type who loved hunting, a man's man and adored by his father. Jacob, the youngest (who came out of the womb holding on to his brother's heel and was thus named *Ya'aqob*, meaning *'heel holder'* or, figuratively, *'supplanter'*), was smooth-skinned, quiet, envious, sneaky and his mother's favourite. Esau was independent and self-assured; Jacob was a 'mummy's boy', with little apparent inclination to think for himself.

When they were grown, an occasion came when Esau arrived home exhausted from the hunt and asked Jacob for some of the red lentil stew that Jacob has just made. Jacob agreed to give him some stew in return for Esau yielding to Jacob the rights of the first-born. Esau thought so little of his birthright in view of his present hunger that he agreed. We can only guess whether Jacob had thought that Esau would be so uncaring of his birthright to yield it so readily but, Esau agreeing, Jacob was quick to make the agreement the subject of a binding oath.

Now, the rules of inheritance of the time required that the male children of a father shared their father's property on his death, but that the fist-born would receive a double portion of the property. In the case of Esau and Jacob, therefore, Jacob now had the right to two thirds of Isaac's property, rather than one third. Since Isaac was an extremely wealthy man, both sons would be very well off.

If the birthright was an important binding legal arrangement relating to family property, the father's blessing was an entirely different matter; it related to the future position of the blessed person in the world. While we might look rather sceptically at the idea of a blessing being either able to foretell the future or to shape a person's future, this was a very powerfully held belief in those times. So Jacob's cheating of Esau out of his blessing (at his mother's instigation), as told in Gen 27:1-40, distressed and infuriated Esau, and he planned to kill Jacob as soon as Isaac died.

Esau married two local women, incorrectly identified in most translations as Hittites. The Hittite peoples were actually located well to the north in what is now called Anatolia and it was many centuries later before the influence of the Hittite Empire stretched down into areas such as Assyria. The actual word translated as Hittite in Gen 26:34 in Hebrew is *Chittiy*, referring to descendants of Heth, the second son of Canaan (whose father had been Ham, son of Noah) and thus one of the Canaanite tribes. This is further illuminated in Gen 27:46, when Rebekah refers to these two as '*bath cheth*' ('daughters of Heth').

There women were a constant source of frustration and annoyance to Isaac and Rebekah, confirming Abraham's wisdom in ensuring that Isaac did not marry into any of the Canaanite tribes. Suspecting Esau's plans to kill Jacob, Rebekah took advantage of the problems that they were having with Esau's wives to convince Isaac to send Jacob to Paddan Aram in north-western Mesopotamia to select a wife from amongst Abraham's relatives. We should not simply think of Rebekah as a schemer—it was to Rebekah that God had revealed that the younger twin was to inherit God's covenant with Abraham, and she would have been fully aware of the need to keep Jacob away from the influences of the Canaanite gods and religious practices. Isaac, too, was fully aware of how important it was that Jacob should be worthy to maintain God's covenant with Abraham to ultimately take possession of the land of Canaan and, as Jacob was preparing to leave, Isaac told him:

> *"May he give the blessing of Abraham to you and to your offspring with you, that you may take possession of the land of your sojournings that God gave to Abraham!" (Gen 28:4)*

And so Jacob, deceiver, sneak and cheat, set out on a journey in which he was to learn what it means to have a personal relationship with God.

17

HOW JACOB BECAME ISRAEL

Jacob's Dream

Jacob's first personal experience of God was when God spoke to him in a dream. On his very first night after leaving the tents of his father and mother, when the reality of what he is about to undertake is beginning to hit him, God makes himself known to Jacob as *'the God of your father Abraham and the God of Isaac'*, not as *'your God'*. To Jacob, God was an unknown quantity. In the dream, God makes a series of promises to Jacob:

> *"The land on which you lie I will give to you and to your offspring. Your offspring shall be like the dust of the earth, and you shall spread abroad to the west and to the east and to the north and to the south, and in you and your offspring shall all the families of the earth be blessed. Behold, I am with you and will keep you wherever you go, and will bring you back to this land. For I will not leave you until I have done what I have promised you." (Gen 28:13-15)*

There were no conditions on God's part.

In return, Jacob makes a vow to God. It is not clear from the original Hebrew whether Jacob responds unconditionally—'*Since* (or *because*) God will be with me ... *then* (therefore) the Lord will be *my* God'—or conditionally—'*If* God will be with me ... so that I return safely ... *then* (at that time) the Lord will be *my* God'. From Jacob's dealings with his uncle, Laban, it seems that the second alternative is most likely and this is the alternative that the majority of Bible translations adopt. Regardless, it is a vow that Jacob meant to keep and did keep—for Jacob, 'God' became '*my* God'.

Thereafter, God reveals himself to Moses and the Children of Israel as '*the God of Abraham, the God of Isaac and the God of Jacob*' (Ex 3:6).

Jacob and Laban

Jacob reached Paddan Aram, found his cousin Laban and fell in love with Laban's daughter Rachel. While initially warm, the relationship between Jacob and Laban became mostly civil but strained as each took advantage of the other. Jacob worked for Laban without payment for seven years in order to marry Rachel, but was tricked by Laban into marrying Leah instead. Laban then agreed to allow Jacob to marry Rachel also, but only on condition that he worked for Laban for another seven years. During those fourteen years, God blessed Laban for Jacob's sake and Laban became rich, but Jacob had gained nothing except eleven sons and a daughter.

Jacob asked Laban to allow him to return to his own home in Beersheba to make his own way in the world. Laban was not happy with the idea; he had been the main beneficiary of God's blessings on Jacob. Instead, he offered to pay Jacob to continue to work for him. Now it was Jacob's turn to take advantage of Laban. He proposed that he be allowed to keep as his own any sheep or goats of Laban's flocks that were born spotted or speckled; any animals born with pure dark coats or white coats would belong to Laban. Laban agreed, but then secretly removed any spotted or speckled animals from his flocks and moved them to a location three day's journey away under his sons' care. Jacob was given animals that were devoid of any spots or speckles.

Jacob seems to have developed an idea that if mating took place in the presence of tree branches that had been stripped of their bark, the resultant lambs and kids would be born speckled or spotted. It seemed to work spectacularly well and Jacob's flocks grew rapidly while Laban's grew more slowly. After six years, Jacob had become a wealthy person in his own right.

We cannot discount the fact that God was keeping his promise to bless Jacob, but we now know that Jacob's success had nothing to do with peeled branches. Modern genetic theory enables us to understand that cross-breeding pure sheep or goats of different colours will produce progeny of which around half will be mixed-colour, meaning that Jacob was able to claim around half of the offspring of Laban's animals for himself. On the other hand, matings between his own speckled/spotted flocks would produce very few progeny that were not speckled or spotted and that thus would have had to be handed back over to Laban.

The reversal of the relative fortunes of Laban and Jacob engendered envy, suspicion and hostility on the part of Laban's sons and Laban's attitude towards Jacob changed for the worse. At this point, Jacob had been part of Laban's household and his employee for twenty years, which placed Jacob in a difficult position—Laban had in effect become his master and could claim ownership over everything that Jacob had. For the first time in twenty years, God appeared to Jacob, probably in a dream, and told him that it was time to return home. So, with the agreement of his wives, Jacob gathered his flocks, his servants and everything he had and fled while Laban was away shearing his sheep. It took three days for Laban to overtake

Jacob who, by that time had made it about three quarters of the way back to Beersheba.

The situation was exacerbated by the fact that Rachel had stolen Laban's household gods before they had fled. Like his father, Bethuel, his grandfather, Nahor, and his great grandfather, Terah, Laban worshipped many gods and, like all similar families, his family had family idols that had been accumulated and passed on from generation to generation. These gods were the only gods that Rachel had ever known and she apparently wanted them close by for protection for her own family (particularly for the protection of her only son, Joseph). To Laban, the loss of his family gods was at least as disastrous as the loss of his daughters and grandchildren. In all innocence of what Rachel had done, Jacob encouraged Laban to search his camp for the idols and declared that the person who took them would die.

The search uncovered nothing because Rachel hid the idols in her camel's saddle and sat on it. She than told the searchers that she was having her period and couldn't stand. Such were the taboos of the time, no living man would touch anything that might be contaminated by menstrual blood and so the saddle was not checked.

Things could have gone very badly for Jacob, with Laban reclaiming his daughters and all Jacob possessed, but God honoured his own promise to Jacob and warned Laban, again probably in a dream, against doing anything to harm Jacob or to prevent his return to his home. In addition, he had apparently been proved wrong in his assertion that Jacob had stolen his household gods. So, instead, a treaty was agreed between them and, having made his goodbyes to his daughters and grandchildren, Laban headed back north.

Later in *Genesis* (Gen 35:4) we learn the fate of those household gods and it was probably slightly prior to that time that the truth of the matter must have come to light so that details of the incident became part of the *Genesis* account.

Jacob Learns to Pray

Jacob now faced what he thought of as a greater threat—his brother, Esau, who had planned to kill him. Jacob was aware that Esau lived well to the south in the land of Seir but he feared that, as soon as Esau learned of his return, his brother would take his revenge. Rather than live in fear, Jacob decided that Esau had to be faced and, if possible, placated. So Jacob decided to do three things: 1) He sent messengers to Esau announcing his return, calling Esau his master and referring to himself as Esau's servant, and asking for Esau's good will; 2) On learning that Esau is on his way with 400 men, he split his people and herds into two groups so that, if Esau attacks one group, the other group might be able to escape; and 3) He prayed.

> *[9] And Jacob said, "O God of my father Abraham and God of my father Isaac, O LORD who said to me, 'Return to your country and to your kindred, that I may do you good,'*
>
> *[10] I am not worthy of the least of all the deeds of steadfast love and all the faithfulness that you have shown to your servant, for with only my staff I crossed this Jordan, and now I have become two camps.*
>
> *[11] Please deliver me from the hand of my brother, from the hand of Esau, for I fear him, that he may come and attack me, the mothers with the children.*
>
> *[12] But you said, 'I will surely do you good, and make your offspring as the sand of the sea, which cannot be numbered for multitude.' (Gen 32:9-12)*

This is the first recorded time that Jacob has prayed in twenty years. No-one has taught him how to pray, but this prayer is a classic example of the prayers that God loves to answer:

- It comes from the heart
- It is based on a real relationship with God (verse 9)
- It recognises the person's own unworthiness and thankfully acknowledges that everything he has comes from God (verse 10)
- It states clearly and simply what God is being asked to do—but not how to do it (verse 11) and
- It is offered in faith that God keeps his promises (verse 12).

In a further attempt to blunt his brother's expected anger, Jacob sent a series of gifts of livestock ahead of him. Jacob had done all that he can think of to handle the situation and that night he sought solitude, perhaps to continue praying.

The Struggle

Genesis 32:24-30 continues the story:

> ^{24} *And Jacob was left alone. And a man wrestled with him until the breaking of the day.*
>
> ^{25} *When the man saw that he did not prevail against Jacob, he touched his hip socket, and Jacob's hip was put out of joint as he wrestled with him.*
>
> ^{26} *Then he said, "Let me go, for the day has broken." But Jacob said, "I will not let you go unless you bless me."*

²⁷ And he said to him, "What is your name?" And he said, "Jacob."

²⁸ Then he said, "Your name shall no longer be called Jacob, but Israel, for you have striven with God and with men, and have prevailed."

²⁹ Then Jacob asked him, "Please tell me your name." But he said, "Why is it that you ask my name?" And there he blessed him.

³⁰ So Jacob called the name of the place Peniel, saying, "For I have seen God face to face, and yet my life has been delivered."

We are not told how this confrontation came about or, initially, who Jacob thought the man was. But it must have become clear that this man was not someone sent either by Laban or Esau to kill him. Neither of them was armed, but the battle was real and intense, and may have lasted for hours. For faithful Jewish and Christian believers, this has always been a highly symbolic story, teaching us to persevere in our prayers for God's help and blessing until God answers. But for all the bruising reality of the encounter, it was also highly symbolic for Jacob; whether the stranger was an angel, as most commentators suggest, or God taking on human form as he apparently had with Abraham on at least one occasion, Jacob began to understand that this was a matter strictly between himself and God. Even after he was disabled, he would not let go until he had received God's blessing.

But the greatest symbolism in the encounter lies in Jacob being asked to speak his own name, *Jacob* (*Heel-grasper, Deceiver, Supplanter*). In giving his name, Jacob was not only confessing his nature (because a person's name reflected their character) he was actually giving the other person power over him (because it was thought that revealing your name to another gave that person power over you). In return, the man gave Jacob a new name—*Israel*. In so doing, the man was effectively telling Jacob that he now had a new nature and a new relationship with God, reflected in his new name.

As is the case for many languages, especially ancient ones, the exact meanings of words are highly contextual. Depending on the context, *Israel* (Hebrew: *Yisra'el*) could mean *struggles, prevails, perseveres with* or *has power (as a prince) with God*. Within the context of the *Genesis* passage, *Perseveres with God* would seem to be the most likely since *Prevails with God* seems to suggest that Jacob was somehow victorious over God.

When Jacob asked the man to reveal his own name, the man answered 'Why do you ask my name?' This answer could be implied either to suggest the meaning 'Isn't it obvious who I am?' or, more likely in the context of the story, that Jacob will not be given power over the man (or God) that possessing the name would provide. Whichever meaning was intended, verse 30 makes it clear that Jacob had no doubts as to who the person was.

Jacob's encounter with Esau is as brotherly as either could have wished, even though it seems that Jacob may still have been wary of Esau's overt friendliness. There is little to make us think that they maintained a close relationship or that they visited each other, but there was clearly no animosity between their respective tribal groups and, on Isaac's death at 180 years old, Jacob and Esau together saw to his burial (Gen 35:28-29).

The Return

For some reason, Jacob did not return immediately to Bethel where, twenty years before, he had made his vow before God. Instead, he moved to Shechem in the northern hill country of Canaan, He lived there for what must have been a few years (perhaps up to a decade) because his sons by Leah must have been at least in their late teens when their sister Dinah is raped by the son of the local tribal leader. The story of that rape and the vengeance of the brothers on the men of Shechem are not relevant to the purposes of this book. However, the ill feeling of the inhabitants of the area towards Jacob's family and a command from God resulted in Jacob at last returning to Bethel.

But Jacob realised that, if *the* God was to be *his* God, he needed to make some changes. His primary task was to ensure that only God was to be worshipped within his household and so ordered all idols and charms held by any members of the group to be yielded up to be buried. Then, probably around thirty years after he had made his promise to God, Jacob finally returned to Bethel, where he built an altar and fulfilled his vow.

In response, God appeared to Jacob, repeating his statement that in future Jacob was to be called Israel and confirming his promise to Jacob that it would be Jacob's descendants who would receive the lands that God had promised to give to the descendants of Abraham and Isaac.

Finally then, Jacob journeyed to re-join his father at Mamre, near Kiriath Arba (later called Hebron). While en route to Mamre, Rachel gave birth to her second son, Benjamin. At this stage, she was probably in her late forties. She died in childbirth.

The Children of Israel

There seem to be some unexplained discrepancies in the ages of Jacob's children, and perhaps even in the order in which they were born. This is best illustrated in the following table in which the children are listed in the Biblical order of their birth.

Child	Mother	Reference
Reuben	Leah	Gen 29:32
Simeon	Leah	Gen 29:33
Levi	Leah	Gen 29:34
Judah	Leah	Gen 29:35
Dan	Bilhah	Gen 30:3-6
Naphtali	Bilhah	Gen 30:7-8
Gad	Zilpah	Gen 30:9-11
Asher	Zilpah	Gen 30:12-13
Issachar	Leah	Gen 30:17-18
Zebulun	Leah	Gen 30:19-20
Dinah	Leah	Gen 30:21
Joseph	Rachel	Gen 30:22-24
Benjamin	Rachel	Gen 35:16-18

According to Gen 30:25, Jacob wanted to leave Laban after Joseph was born and that this marked the end of the seven extra years after Jacob had worked for Rachel and thus the end of Jacob's first seven years of marriage. If we accept that Dinah may have been born 'some time later' (Gen 30:21), this still means that Leah conceived and gave birth to six sons in a period of seven years. To complicate matters, Leah gave her maidservant Zilpah to Jacob after Leah believed that she had ceased child-bearing, implying that, for an extended period of at least 18 months during those seven years, Leah was not pregnant.

In all probability, Rachel gave Bilhah to Jacob only a year or two into their marriage, so Bilhah would have been pregnant with her two boys at times when Leah and Zilpar were also pregnant and Rachel could have been pregnant at the same time that Leah was pregnant with Zebulun. Consequently, the birth order between sons of the one mother is correctly stated, but the chronological order of children of different mothers was probably very different to that shown above.

This still leaves us with another problem that is not so easily resolved—Gen 37:3 refers to Joseph as being Jacob's favourite because he had been born to Jacob 'in his old age'. But if the first eleven sons were all born in a seven-year period, this does not make sense. We can only assume that the records concerning Jacob and his children from which the *Genesis* account was constructed were piecemeal and poorly prepared, maintained or translated.

Benjamin was certainly a son of Jacob's old age, since it is probable that he was born more than fifteen years after Joseph.

18

JOSEPH-FROM SLAVE TO SAVIOUR

Joseph the Dreamer

We are first introduced to Joseph as a seventeen year old, probably not long after Jacob, now called Israel, returned to Kiriath Arba with his family. We are told that Joseph brought a bad report about his older brothers back to Israel and earned their hatred, made worse by the fact that he was their father's favourite. As previously noted, Israel's love for Joseph is said to be because he was 'the son of his old age' but, more likely it was because he was Rachel's son (at this stage, Rachel's youngest son, Benjamin, was probably still a babe in arms).

Joseph had dreams that implied that one day his brothers would bow down to him, which offended his brothers even further. Isaac, on the other hand, was aware that these dreams may very well symbolise part of the fulfilment of God's covenantal promise. In the brothers' position, we too would probably have regarded Joseph as a 'tattle-tail' and someone 'too big for his boots'. While later events proved otherwise, it was a necessary part of God's plan for the future of the Children of Israel.

It was probably not long after Joseph told his father and brothers of his second dream that his ten older brothers were sent with the family's flocks to find fresh grazing land in the area near Shechem, about twenty miles north of Bethel. It may be that all ten of the brothers went, not because of the size of the flocks, large as these might have been, but because the inhabitants of that area were still resentful of Leah's sons' revenge on the men of Shechem and it was better to keep the brothers as a single strong defensive force. It also provides an explanation as to why Israel sent out Joseph weeks later to check on their well-being.

Joseph finally located his brothers near Dothan, some thirteen miles north of Shechem. The brothers saw this as a great opportunity to get rid of Joseph and planned to kill him, first putting him into one of the deep, dry cisterns in the area. As the eldest child and the one responsible for the whole party, Reuben wanted to save Joseph, but while he was absent, the others sold Joseph as a slave to a passing group of Midianite merchants who were on their way to Egypt. When Reuben returned, he was faced with a *fait accompli* and was forced to go along with a plan to slaughter one of the flock, stain Joseph's richly embroidered robe with its blood and take the robe back to Israel so that he would think that Joseph had been killed by a wild animal (at that time and for centuries after, many savage beasts, including lions, were native to the region).

Joseph in Egypt

Genesis tells us that Joseph was sold by the Midianites to Potiphar, Pharaoh's official chief bodyguard (Hebrew: *Par'oh cariyc sar tabbach*). The Hebrew word *tabbach* literally means *butcher* and, by extension can refer to an *executioner*, *cook* or *guard*. It is possible that Potiphar was the Pharaoh's chief cook, but the context seems to imply the more important role of bodyguard (by comparison, the head of the prison where Pharaoh's prisoners were kept was referred to as *sar bayith cohar* (chief of the house surrounded by walls). The word *cariyc* (official) is not part of his title, implying that the position was not a pharaohnic appointment.

Joseph didn't despair or sulk or get angry over the injustice that had been done to him. He had an absolute trust that God would look after him, in whatever circumstances Joseph found himself, and so he set out to be a person of integrity and to please both God and man.

In Genesis 39:2-6, we learn that:

> *² The LORD was with Joseph, and he became a successful man, and he was in the house of his Egyptian master.*
>
> *³ His master saw that the LORD was with him and that the LORD caused all that he did to succeed in his hands.*
>
> *⁴ So Joseph found favor in his sight and attended him, and he made him overseer of his house and put him in charge of all that he had.*
>
> *⁵ From the time that he made him overseer in his house and over all that he had, the LORD blessed the Egyptian's house for Joseph's sake; the blessing of the LORD was on all that he had, in house and field.*
>
> *⁶ So he left all that he had in Joseph's charge, and because of him he had no concern about anything but the food he ate. Now Joseph was handsome in form and appearance.*

Potiphar's position probably kept him at the palace every day and for the greater part of each day, so it is not surprising that his wife cast a wandering eye over Joseph's '*handsome form and appearance*'. It says a lot for Joseph's integrity and his faithfulness to God that he consistently resists her approaches. Finally, one day he is cornered by her and, in his efforts to escape, left his garment (Heb. *beged*) behind in her hand. All Egyptian men before about 1400 BC wore a loincloth or 'skirt' (called a *shendyt*) or, as formal wear, a one-piece short-sleeved tunic that covered the top part of their body as well as their thighs. After that time, some form of shirt became popular. But in Joseph's time servants, even chief servants, were only ever likely to wear a *shendyt*, so Joseph almost certainly fled naked.

True to the old saying about a woman scorned, Potiphar's wife accused Joseph of attempted rape and he was referred to as 'that Hebrew', the only person other than Abraham to be called a Hebrew in *Genesis*, and this time the term was used pejoratively. Potiphar's position was such that he could almost certainly have had Joseph executed, but God had plans for Joseph and, for some reason known only to Potiphar, Joseph was simply thrown into the prison specifically designated for prisoners of the pharaoh.

Joseph did not know why God allowed bad things to happen to him. Being firstly sold as a slave and then imprisoned under false allegations must have been both confusing and distressing, but he never allowed these things to change the kind of person he was. He was a child of Israel, an inheritor of God's promises. He would continue to be a person of integrity and one wholly determined to serve his God whatever the cost. Like King David centuries later, his attitude towards God seems to have been '*My times are in your hand*' (Psa 31:15).

In prison, he conducted himself honourably and respectfully and won the complete trust of the head of the prison, so much so that Joseph effectively became the prison's deputy head and chief administrator.

Joseph was not just a dreamer—under the guidance of God, he also had the ability to interpret dreams. In those days, unusual dreams were regarded as messages from the gods and every royal court had officials wise in the interpretation of dreams. When two of the pharaoh's officials, the chief cupbearer and the chief baker, were imprisoned after in some way displeasing Pharaoh, each had dreams that disturbed them. By this time, Joseph had befriended both men and offered to interpret their dreams. He correctly interpreted both dreams—in which the chief baker was to be executed and the chief cupbearer was to be restored to his position - asking only that the cupbearer would mention Joseph's case with the pharaoh to gain Joseph's release. But the cupbearer forgot about Joseph for two more years, by which time Joseph had been in Egypt for about twelve years, probably at least half of that time in prison.

At the end of that time, *Genesis* tells us that the pharaoh had two strangely similar dreams in the one night and awoke very troubled. All the wise men and magicians were called to provide an interpretation of the dreams, but no-one could. It was then that the cupbearer remembered Joseph and told pharaoh about his own and the baker's dreams and how Joseph had correctly interpreted them. The pharaoh immediately sent for Joseph, who was then shaved (because facial hair was unacceptable in a pharaoh's presence), freshly dressed and brought to the palace. Genesis 41:15-16 tells us:

> [15] *And Pharaoh said to Joseph, "I have had a dream, and there is no one who can interpret it. I have heard it said of you that when you hear a dream you can interpret it."*
>
> [16] *Joseph answered Pharaoh, "It is not in me; God will give Pharaoh a favorable answer."*

Of course, Joseph was able to interpret the dreams, forecasting seven very bountiful years of harvests to be followed by seven disastrous years in which the Nile floods would fail and famine would ravage the land. He also suggested that Pharaoh appoint a wise official to oversee commissioners whose job it would be during the seven good years to gather and store twenty percent of the harvests for use in the seven years of famine. Perhaps unsurprisingly, Pharaoh appointed Joseph to the post of vizier, making him second only to Pharaoh in the land of Egypt. He continued to hold that position throughout the seven years of plenty and the seven years of drought and there is every reason to believe he would have continued in the position at least until the death of the pharaoh that appointed him. He lived to the age of 110 years and was vizier from the age of thirty.

No Egyptian records have been found to date that mention Joseph, either by that name or by his Egyptian name of *Zaphenath-Paneah*. In many respects this could be understandable if it were not for the apparent importance of Joseph's purported position. However, there are large blank periods in Egypt's history of the time and more records are being uncovered as archaeological work continues in Egypt to the present day.

In 1 Kings 6:1 we are told that the Exodus took place 480 years before the fourth year of King Solomon's reign over Israel, which was 966 BC. This places the date of the Exodus at 1446 BC. Exodus 12:40-41 tells us that '*The time that the people of Israel lived in Egypt was* 430 years. *At the end of 430 years, on that very day, all the hosts of the Lord went out from the land of Egypt*'. This gives us a date for Israel's arrival in Egypt of around 1876 BC. By that time, Joseph had been vizier for about nine or ten years and thus places Joseph under the reigns of Twelfth Dynasty pharaohs Sesostris II (1897-1878 BC) and his son, Sesostris III (1878-1860 BC). Sesostris III was the most powerful of the Middle Kingdom pharaohs and the success of his reign could possibly have arisen from Joseph's wide-scale purchase of private lands (Gen 47:18-26) for Pharaoh in exchange for food and the income that was produced from leasing the land back to the original owners.

The 430 years the Hebrews lived in Egypt would then have bridged the period from the late Middle Kingdom, through the turbulent Second Intermediate Period (1802-1550 BC) into the beginnings of the New Kingdom (1550-1077 BC). During the latter part of the Second Intermediate Period, the Hyksos people seized power over most of Egypt, ruling from around 1674-1535 BC with the centre of their administration located in the Nile Delta country.

Until recently, archaeologists and historians has assumed that the Hyksos had invaded from Asia down through the Fertile Crescent into Egypt at that time. However, in2020 it was announced that scientists studying the teeth of Hyksos skeletons were able to show that the Hyksos had already been settled in the Nile Delta region for some generations prior to their takeover of the country. This gives rise to the possibility that the Hebrews were allocated land in Goshen (which was located in the Nile Delta region) because this was an area where 'immigrants' were encouraged to live.

Around 1535 BC, the Egyptians rose up against the hated Hyksos rulers and succeeded in driving them out to begin the New Kingdom Period. Middle Kingdom history was probably largely forgotten by the time of the New Kingdom and the fact that the Hebrews had lived cheek-by-jowl with the Hyksos people in Goshen probably resulted in the Hebrews being associated with them in the minds of the Egyptians who consequently persecuted them.

In an alternative version of possible Egyptian history, some scholars have noted that a city named Rameses is found in Ex 1:11 and suggest that the pharaoh at the time of the Exodus was Rameses II ('Rameses the Great'). This would place the Exodus at a much later date, around 1290 BC. If this were the case, it would change the date of Israel's settlement in Egypt to around 1720 BC, during the Thirteenth Dynasty within the Second Intermediate Period. This seems to be too unstable a period in which the events described in *Genesis* could have occurred. In addition, Gen 47:11 states that Joseph gave his father and his brothers land in the '*district of Rameses*'. It could not possibly have been called that at the time and the most likely explanation is that both of these passages were subject to editing by people copying the documents at a later date (there are a number of such examples of editing in the Bible).

I have ignored the bulk of the story of Joseph's treatment of his brothers as being of little scientific interest, but it is worthwhile noting that the ordeals he put them through on both trips were not done from spite—he could have done far worse. Instead, they showed a fine sense of psychology in serving two purposes. Firstly, it taught them what it is really like to be treated unjustly and to find themselves powerless, subject to the whims of others. Perhaps they would better understand the trauma that Joseph felt when sold into slavery in a new land, unable even to speak the language. Secondly, his insistence that Benjamin must come on their second trip and the anguish that this caused their father had to remind them of the deep and lasting anguish that their stealing and selling of Joseph had inflicted on him years before and continued to weigh him down after all these years. Joseph could have simply overlooked what they had done and the damage it had caused and no lesson would have been learned; but instead, they were forced to confront all the pain that their earlier actions had brought about. This is a lesson we must all understand—it is only when we comprehend the consequences to others and to ourselves of the hurtful or criminal things that we have done that we can truly repent.

When Israel died, Joseph's brothers were afraid that Joseph would have them killed. So they sent a message to Joseph saying that Israel had asked Joseph to forgive them. His response, in Gen 50:18-21 is a fitting climax to *Genesis*:

[18] His brothers also came and fell down before him and said, "Behold, we are your servants."

[19] But Joseph said to them, "Do not fear, for am I in the place of God?

[20] As for you, you meant evil against me, but God meant it for good, to bring it about that many people should be kept alive, as they are today.

[21] So do not fear; I will provide for you and your little ones." Thus he comforted them and spoke kindly to them.

Joseph realised that all that happened to him was part of God's specific plan for saving his people. Throughout all of his troubles, God was ensuring that Joseph would be in the right place at the right time to fulfil God's purpose for him.

My hope and prayer is that every reader will live faithfully in God's sight so that they, too, might become part of God's great plan and, in particular, might find God's specific plan for his or her life.

CONCLUSION

Now we have gone through the *Book of Genesis*, what can we say about it? For over 3500 years this book has provided Jewish theologians with an explanation for the origins of the universe and of life. For 2000 years it has done the same for Christians. But is it true? And is it relevant?

Truth and Consequences

There is no doubt that the science of Genesis is simplistic, but it was written by persons whose scientific understanding was simplistic and initially was written for people of simple understanding. Its two major tenets are that all of creation and all life came into existence by the will and power of a single Creator God and that this creation was for the purpose of raising up a race of beings who could actively and willingly enter into a personal relationship with their Creator.

Science aside, the sheer difficulty of maintaining coherency of stories through generations of illiteracy and innumeracy, then transcribing these stories through who-knows-how-many copies and edits is immense. What is truly remarkable is that, despite undoubted errors and omissions, the individuals and events fairly accurately reflect the languages, attitudes and mores of the times as we have been able to establish them through modern archaeological and linguistic research. On this basis alone, there is good reason to accept the historical veracity of many of these characters and events, even if many details have clearly been exaggerated or mythologised.

If its science is simplistic, *Genesis'* symbolism is complex and significant. As we have considered back in Chapter 3, if there is a Creator God, he has no relevance to us unless we are relevant to him. On this basis, we would expect that such a God would reveal himself and, in *Genesis*, we see the beginning of such a revelation. But this revelation is not painted across the sky in way we might expect if it was his purpose to impose a blanket worship of himself out of awe, fear or obligation. Instead, we are shown a God who reveals himself to those who are open to him. Moreover, he makes clear that any relationship with him must be based on our willing submission to him.

Genesis also makes it clear that God is just not an observer; he is actively involved in this world, both in the working out of his long-term plan and also in response to our intercessions on behalf of ourselves and others. Every book in the Old and New Testaments confirm this and countless Christians down through the last 2000 years have proved how accessible and responsive God is to our prayers.

The Greatest Experiment

Genesis takes us from God's revelation of himself to one man, continuing through that man's descendants to a tribe that fully acknowledged and obeyed God as the One True God and which, in the course of time, became a nation under God. Through that nation, God himself appeared in the person of his son, Jesus, to share our lives and experience our pleasures and pains. God, as the Great Scientist, had created the greatest experiment of all time, entering into and sharing with his own creation. The outcome of which was that, in turn, any who are willing to submit to him might enter into and share in his eternal presence.

I hope, dear reader, that this has been or will be your experience.

www.ingramcontent.com/pod-product-compliance
Lightning Source LLC
Chambersburg PA
CBHW060131100426
42744CB00007B/753